Lecture Notes in Mathematics

Edited by A. Dold and B. Eckmann

Subseries: Fondazione C.I.M.E., Firenze
Adviser: Roberto Conti

996

Invariant Theory

Proceedings of the 1st 1982 Session of the
Centro Internazionale Matematico Estivo (C.I.M.E.)
Held at Montecatini, Italy, June 10–18, 1982

Edited by F. Gherardelli

Springer-Verlag
Berlin Heidelberg New York Tokyo 1983

Editor

Francesco Gherardelli
Istituto Matematico U. Dini, Università degli Studi
Viale Morgagni 67 A, 50134 Firenze, Italy

AMS Subject Classifications (1980): 14 D 20, 14 D 25

ISBN 3-540-12319-9 Springer-Verlag Berlin Heidelberg New York Tokyo
ISBN 0-387-12319-9 Springer-Verlag New York Heidelberg Berlin Tokyo

2146/3140-543210

C.I.M.E. Session on Invariant Theory

List of Participants

B. Ådlandsvik, Matematisk Institutt, Allégt. 55, 5000 Bergen, Norway

A. Albano, Istituto di Geometria "C. Segre", Via Principe Amedeo 8, 10123 Torino

G. Almkvist, PL 500, 24300 Höör, Sweden

L. Amodei, Istituto Matematico Università, Via F. Buonarroti 2, 56100 Pisa

F. Arnold, Willy-Andreas-Allee 5, 7500 Karlsruhe, West Germany

E. Ballico, Scuola Normale Superiore, Piazza dei Cavalieri 7, 56100 Pisa

R. Benedetti, Istituto Matematico "L. Tonelli", Via F. Buonarroti 2, 56100 Pisa

J. Bertin, 18 rue du Sénéchal, 31000 Toulouse, France

C. Blondel, UER de Math. 45-55, Univ. Paris VII, 2 Pl. Jussieu, 75221 Paris, France

J. F. Boutot, 15 rue Erard, 75012 Paris, France

M. Brion, E.N.S., 45 rue d'Ulm, 75005 Paris, France

F. Catanese, Istituto Matematico Università, Via F. Buonarroti 2, 56100 Pisa

A. Collino, Istituto di Geometria, Università, Via Principe Amedeo 8, 10123 Torino

P. Cragnolini, Istituto Matematico Università, Via F. Buonarroti 2, 56100 Pisa

M. Dale, Matematisk Institutt, Allégt. 55, 5000 Bergen, Norway

C. De Concini, Istituto Matematico Università, Via F. Buonarroti 2, 56100 Pisa

A. Del Centina, Istituto Matematico Università, Viale Morgagni 67/A, 50134 Firenze

G. D'Este, Istituto di Algebra e Geometria, Via Belzoni 7, 35100 Padova

D. Dikranjan, Istituto Matematico Università, Via F. Buonarroti 2, 56100 Pisa

G. Elencwajg, IMSP-Mathématiques, Parc Valrose, 06034 Nice-Cedex, France

J. Eschgfäller, Istituto Matematico Università, Via Machiavelli 35, 44100 Ferrara

P. Gerardin, UER de Mathématiques, Université Paris VII, 2 Place Jussieu, 75231 Paris-Cedex, France

F. Gherardelli, Istituto Matematico Università, Viale Morgagni 67/A, 50134 Firenze

P. Gianni, Dipartimento di Matematica, Via F. Buonarroti 2, 56100 Pisa

D. Gieseker, Department of Mathematics, UCLA, Los Angeles, Cal. 90024, USA

A. Gimigliano, Viale della Repubblica 85, 50019 Sesto Fiorentino, Firenze

K. Girstmair, Institut für Mathematik, Universität Innsbruck, Innrain 52, A-6020 Innsbruck, Austria

A. Helversen-Pasotto, Laboratoire associé du CNRS n. 168, Départment de Mathématique, IMSP, Université de Nice, Parc Valrose, 06034 Nice-Cedex, France

T. Johnsen, Lallakroken 8 C, Oslo 2, Norway

V. Kac, Department of Mathematics, MIT, Cambridge, Mass. 02139, USA

M. Laglasse-Decauwert, USMG Institut Fourier, Mathématiques Pures, B.P. 116, 38402 Saint Martin D'Heres, France

A. Lanteri, Istituto Matematico "F. Enriques", Via C. Saldini 50, 20133 Milano

A. Lascoux, L.I.T.P., UER Maths Paris VII, 2 Place Jussieu, 75251 Paris Cedex 05, France.

D. Luna, 12 Rue Fracy, 38700 La Tronche, France

V. Mehta, Department of Mathematics, The University of Bombay, Bombay, India

M. Meschiari, Via Baraldi 12, 41100 Modena

D. Montanari, Via Asmara 38, 00199 Roma

J. Oesterlé, E.N.S., 45 rue d'Ulm, 75005 Paris, France

P. Oliverio, Scuola Normale Superiore, Piazza dei Cavalieri 7, 56100 Pisa

M. Palleschi, Via Bergognone 27, 20144 Milano

F. Pauer, Institut für Mathematik, Universität Innsbruck, Innrain 52, A-6020 Innsbruck, Austria

E. Previato, Department of Mathematics, Harvard University, Cambridge,Mass.02138, USA

M. Roberts, 9 Pelham Grove, Liverpool 17, England

N. Rodinò, Via di Vacciano 87, 50015 Grassina, Firenze

S. Rosset, School of Mathematics, University of Tel-Aviv, Tel-Avis, Israele

P. Salmon, Via Rodi 14/9, 16145 Genova

W. K. Seiler, Mathematisches Institut II, Englerstrasse 2, 7500 Karlsruhe 1, West Germany

M. Seppälä, Fakultat für Mathematik, Universität Bielefeld, BRD-4800 Bielefeld 1, Germania Occ.

R. Smith, Univ. of Georgia, Graduate Students Res. Center, Athens,Ga 30602, USA

C. Traverso, Dipartimento di Matematica, Via F. Buonarroti 2, 56100 Pisa

C. Turrini, Istituto Matematico "F. Enriques", Via C. Saldini 50, 20133 Milano

L. Verdi, Istituto Matematico Università, Viale Morgagni 67/A, 50134 Firenze

A. Verra, Via Assarotti 16, 10122 Torino

G. Weill, 300 E 33rd Street, New York, N.Y. 10016, USA

INDEX

COMPLETE SYMMETRIC VARIETIES

by

C. De Concini and C. Procesi

Università di Roma II — Università di Roma

Nur der Philister schwärmt für absolute Symmetrie
H. Seidel, ges. w. 1,70 *

INTRODUCTION

In the study of enumerative problems on plane conics the following variety has been extensively studied ([6],[7],[15],[17],[18],[19],[20],[23],[25]).

We consider pairs (C,C') where C is a non degenerate conic and C' its dual and call X the closure of this correspondence in the variety of pairs of conics in $\mathbb{P}^2$ and $\check{\mathbb{P}}^2$.

On this variety acts naturally the projective group of the plane and one can see that X decomposes into 4 orbits: X_o open in X; X_1, X_2 of codimension 1 and $X_3 = \bar{X}_1 \cap \bar{X}_2$ of codimension 2. All orbit closures are smooth and the intersection of $\bar{X}_1$ with $\bar{X}_2$ is transversal. This theory has been extended to higher dimensional quadrics ([1],[15],[17],[21]) and also carried out in the similar example of collineations ([16]).

The renewed interest in enumerative geometry (see e.g. [11]) has brought back some interest in this class of varieties ([22],[5] , cf.§6).

In this paper we will study closely a general class of varieties, including the previous examples, which have a significance for enumerative problems.

Let $\bar{G}$ be a semisimple adjoint group, $\sigma: \bar{G} \to \bar{G}$ an automorphism of order 2 and $\bar{H} = \bar{G}^\sigma$. We construct a canonical variety X with an action of $\bar{G}$ such that

1) X has an open orbit isomorphic to $\bar{G}/\bar{H}$
2) X is smooth with finitely many $\bar{G}$ orbits
3) The orbit closures are all smooth
4) There is a 1-1 correspondence between the set of orbit closures and the family of subsets of a set I_ℓ with ℓ elements. If $J \subseteq I_\ell$ we denote by S_J the corresponding orbit closure
5) We have $S_I \cap S_J = S_{I \cup J}$ and codim S_I = card I

* We thank the "Lessico intellettuale europeo" for supplying the quotation.

6) Each S_I is the transversal complete intersection of the $S_{\{u\}}$, $u \in I$

7) For each S_I we have a $\bar{G}$ equivariant fibration $\pi_I : S_I \to G/P_I$ with P_I a parabolic subgroup with semisimple Levi factor L, σ stable, and the fiber of π_I is the canonical projective variety associated to L and $\sigma|L$

Using results of Bialynicki Birula [2] we give a paving of X by affine spaces and compute its Picard group. We describe the positive line bundles on X and their cohomology in a fashion similar to that of "Flag varieties".

Next we give a precise algorithm which allows to compute the so called characteristic numbers of basic conditions (in the classical terminology) in all cases. The computation can be carried out mechanically although it is very lengthy.

As an example we give the classical application due to H.Schubert [14] for space quadrics and compute the number of quadrics tangent to nine quadrics in general position.

We should now make three final remarks. First of all our method has been strongly influenced by the work of Semple [15], we have in fact interpeted his construction in the language of algebraic groups. The second point will be taken in a continuation of this work.Briefly we should say that a general theory of group embeddings due to Luna and Vust [13] has been used by Vust to classify all projective equivariant embeddings of a symmetric variety of adjoint type and in particular the ones which have the property that each orbit closure is smooth. We call such embeddings wonderful. It has been shown by Vust that such embeddings are all obtained in most cases from our variety X by successive blow ups, followed by a suitable contraction.

This is the reason why we sometimes refer to X as the minimal compactification, in fact it is minimal only among this special class.

The study of the limit provariety obtained in this way is the clue for a general understanding of enumerative questions on symmetric varieties as we plan to show elsewhere.

Finally we have restricted our analysis to characteristic 0 for simplicity. Many of our results are valid in all characteristics (with the possible exception of 2) and some should have a suitable characteristic free analogue. Hopefully an analysis of this theory may have same applications to representation theory also in positive characteristic.

The first named author wishes to thank the Tata institute of Fundamental research and the C.N.R. for partial financial support during the course of this research. Special thank go to the C.I.M.E.

which allowed him to lecture on the material of this paper at the meeting on the "Theory of Invariants" held in Montecatini in the period June 10-18, 1982.

The second named author aknowledges partial support from Brandeis University and grants from N.S.F. and C.N.R. during different periods of the development of this research.

1. PRELIMINARIES

In this section we collect a few more or less well known facts.

1.1. Let G be a semisimple simply connected algebraic group over the complex numbers. Let $\sigma: G \to G$ be an automorphism of order 2 and $H=G^{\sigma}$ the subgroup of G of the elements fixed under σ. The homogeneous space G/H is by definition a symmetric variety and more generally, if G' is a quotient of G by a (finite) σ stable subgroup of the center of G, the corresponding G'/H' will again be a symmetric variety.

Let $\underline{g}$, $\underline{h}$ denote the Lie algebras of G, H respectively. σ induces an automorphism of order 2 in $\underline{g}$ which will again be denoted by σ and $\underline{h}$ is exactly the +1 eigenspace of σ.

We recall a well known fact:

PROPOSITION. Every σ-stable torus in G is contained in a maximal torus of G which is σ stable.

If T is a σ stable torus and $\underline{t}$ its Lie algebra, we can decompose $\underline{t}$ as $\underline{t} = \underline{t}_0 \oplus \underline{t}_1$ according to the eigenvalues +1, -1 of σ. $\underline{t}_0$ is the Lie algebra of the torus $T_0 = T^{\sigma}$ while $\underline{t}_1$ is the Lie algebra of the torus $T_1 = \{t \in T | t^{\sigma} = t^{-1}\}$ such a torus is called anisotropic. The natural mapping $T_0 \times T_1 \to T$ is an isogeny, it is not necessarily an isomorphism since the character group of T need not decompose under σ into the sum of the subgroups relative to the eigenvalues ±1. We indicate still by σ the induced mapping on $\underline{t}^*$ and can easily verify in case T is a maximal torus and $\Phi \subseteq \underline{t}^*$ the root system:

i) If $\underline{t} \oplus \sum_{\alpha\in\Phi} \underline{g}_\alpha$ is the root space decomposition of $\underline{g}$ then

$\sigma(\underline{g}_\alpha) = \underline{g}_{\alpha\sigma}$, hence $\sigma(\Phi) = \Phi$.

(ii) σ preserves the Killing form.

We want now to choose among all possible σ stable tori one for which dim T_1 is maximal and call this dimension the rank of G/H, indicated by ℓ.

1.2. Having fixed T and so the root system Φ we proceed now to fix the positive roots in a compatible way.

LEMMA. One can choose the set Φ^+ of positive roots in such a way that: If $\alpha \in \Phi^+$ and $\alpha \not\equiv 0$ on $\underline{t}_1$ then $\alpha^\sigma \in \Phi^-$.

PROOF. Decompose $\underline{t}^* = \underline{t}_o^* \oplus \underline{t}_1^*$; every root α is then written $\alpha = \alpha_o + \alpha_1$ and $\alpha^\sigma = \alpha_o - \alpha_1$. Choose two R-linear forms ϕ_o and ϕ_1 on $\underline{t}_o^*$ and $\underline{t}_1^*$ such that ϕ_o and ϕ_1 are non zero on the non zero components of the roots. We can replace ϕ_1 by a multiple if necessary so that, if $\alpha = \alpha_o + \alpha_1$ and $\alpha_1 \neq 0$ we have $|\phi_1(\alpha_1)| > |\phi_o(\alpha_o)|$. Consider now the R-linear form $\phi = \phi_o \oplus \phi_1$, we have that $\phi(\alpha) \neq 0$ for every root α; moreover if $\alpha \not\equiv 0$ on t_1, i.e. $\alpha = \alpha_o + \alpha_1$ with $\alpha_1 \neq 0$ the sign of $\phi(\alpha)$ equals the sign of $\phi_1(\alpha_1)$. Thus, setting $\Phi^+ = \{\alpha \in \Phi | \phi(\alpha) > 0\}$ we have the required choice of positive roots. Let us use the following notations

$$\Phi_o = \{\alpha \in \Phi |\ \alpha | t_1 = 0\}, \quad \Phi_1 = \Phi - \Phi_o .$$

Clearly $\Phi_o = \{\alpha \in \Phi | \alpha^\sigma = \alpha\}$ while by the previous lemma σ interchanges Φ_1^+ with Φ_1^-.

Having fixed Φ^+ as in the above lemma we denote by $B \subset G$ the corresponding Borel subgroup and by B^- its opposite Borel subgroup.

1.3. It is now easy to describe the Lie algebra $\underline{h}$ in terms of the root decomposition. We have already noticed that $\sigma(\underline{g}_\alpha) = \underline{g}_{\alpha^\sigma}$.

LEMMA. If $\alpha \in \Phi_o$, σ is the identity on g_α.

PROOF. Let x_α, y_α, h_α be the standard sl_2 triple associated to α. Since $\alpha^\sigma = \alpha$ we have $\sigma(h_\alpha) = h_\alpha$. On the other hand since $\sigma(g_{\pm\alpha}) = g_{\pm\alpha}$ we have $\sigma(x_\alpha) = \pm x_\alpha$. Now if $\sigma(x_\alpha) = -x_\alpha$ we must have also $\sigma(y_\alpha) = -y_\alpha$ since $h_\alpha = [x_\alpha, y_\alpha]$. Now if we consider any element $s \in \underline{t}_1$ we have $[x_\alpha, s] = [y_\alpha, s] = 0$ since α vanishes on $\underline{t}_1$ by hypothesis. This implies, setting $t = x_1 + y_1$, that $\underline{t}_1 + Ct$ is a Toral subalgebra on which σ acts as -1. Since we can enlarge this to a maximal Toral subalgebra, we contradict the choice of T maximizing the dimension of T_1.

PROPOSITION. $\underline{h} = \underline{t}_o + \sum_{\alpha \in \Phi_o} g_\alpha + \sum_{\alpha \in \Phi_1} C(x_\alpha + \sigma(x_\alpha))$.

PROOF. Trivial from the previous lemma.

We may express a consequence of this, the so called Iwasawa decomposition: The subspace $\underline{t}_1 + \sum_{\alpha \in \Phi_1^+} Cx_\alpha$ is a complement to $\underline{h}$ and so it projects isomorphically onto the tangent space of G/H at H, in particular since Lie B $\supset t_1 + \sum_{\alpha \in \Phi_1^+} \mathbb{C}x_\alpha$, BH $\subset$ G is dense in G.

COROLLARY. $\dim G/H = \dim \underline{t}_1 + 1/2|\Phi_1|$.

1.4. If $\Gamma \subset \Phi_+$ is the set of simple roots, let us denote $\Gamma_o = \Gamma \cap \Phi_o$, $\Gamma_1 = \Gamma \cap \Phi_1$ explicitely:

$$\Gamma_o = \{\beta_1,\dots,\beta_k\}; \qquad \Gamma_1 = \{\alpha_1,\dots,\alpha_j\}.$$

LEMMA. For every $\alpha_i \in \Gamma_1$ we have that α_i^σ is of the form $-\alpha_k - \Sigma n_{ij}\beta_j$ for some $\alpha_k \in \Gamma_1$ and some non negative integers n_{ij}. Moreover, $\alpha_k^\sigma = -\alpha_i - \Sigma n_{ij}\beta_j$.

PROOF. By Lemma 1.2 we know that $\alpha_i^\sigma \in \Phi^-$ hence we can write $\alpha_i^\sigma = -(\Sigma m_{ik}\alpha_k + \Sigma n_{ij}\beta_j)$ where m_{ik} , n_{ij} are non negative integers.Thus $\alpha_i = \alpha_i^{\sigma\sigma} = \sum_k m_{ik}(\sum_t m_{kt}\alpha_t) + \Sigma m_{ik}\Sigma n_{kj}\beta_j - \Sigma n_{ij}\beta_j$. Since the simple roots are a basis of the root lattice we must have in particular $\Sigma m_{ik}m_{kt} = 0$ for $t \neq i$ and $\Sigma m_{ik}m_{ki} = 1$. Since the m_{ij}'s are non negative integers it follows that only one m_{ik} is non zero and equal to 1 and the m_{ki} is also equal to 1.

Now consider the fundamental weights. Since they form a dual basis of the simple coroots we also divide them:

$$\omega_1,\dots,\omega_j, \qquad \zeta_1,\dots,\zeta_k \qquad \text{where:}$$

$$(\omega_i , \check\beta_j) = 0, \qquad (\omega_i , \check\alpha_j) =: \delta_j^i$$ and similarly for the ζ_j's.

Since σ preserves the Killing form we have:

$$(\omega_i^\sigma , \check\beta_j^\sigma) = (\omega_i^\sigma , \check\beta_j) = 0$$

$$\delta_j^i = (\omega_j^\sigma , \check\alpha_i^\sigma) = (\omega^\sigma , \frac{2}{(\alpha_i , \alpha_i)}(-\alpha_k - \Sigma n_{ij}\beta_j))$$

$$= -(\omega_j^\sigma , \frac{2\alpha_k}{(\alpha_i , \alpha_i)}) = \frac{(\alpha_k , \alpha_k)}{(\alpha_i , \alpha_i)}(\omega_j^\sigma , \check\alpha_k)$$

We deduce that

$$\omega_i^\sigma = -\frac{(\alpha_k , \alpha_k)}{(\alpha_i , \alpha_i)}\omega_k.$$

Now ω_i^σ must be in the weight lattice so $\frac{(\alpha_k\ ,\ \alpha_k)}{(\alpha_i\ ,\ \alpha_i)}$ is an integer. Reversing the role of i and k we set that it must be 1 so

$$\omega_i^\sigma = -\omega_k .$$

We can summarize this by saying that we have a permutation $\tilde{\sigma}$ of order 2 in the indices 1,2,...,j such that $\omega_i^\sigma = -\omega_{\tilde{\sigma}(i)}$.

DEFINITION. A dominant weight is special if it is of the form $\Sigma n_i\omega_i$ with $n_i = n_{\tilde{\sigma}(i)}$. A special weight is regular if $n_i \neq 0$ for all i.

Thus we have that a weight λ is special iff $\lambda^\sigma = -\lambda$.

1.5.

LEMMA. Let λ be a dominant weight and let V_λ the corresponding irreducible representation of G with highest weight λ. Then if V_λ^H denotes the subspace of V_λ of H-invariant vectors dim $V_\lambda^H \leq 1$ and if $V_\lambda^H \neq 0$ λ is a special weight.

PROOF. Recall that $BH \subset G$ is dense in G so that H has a dense orbit in G/B. Also $V_\lambda \underset{G}{\cong} H^O(G/B,L)$ for a suitable line bundle L on G/B. So if $s_1, s_2 \in V_\lambda^H - \{0\}$, we have that $\frac{s_1}{s_2}$ is a meromorphic function on G/B constant on the dense H orbit, hence s_1 is a multiple of s_2 and our first claim follows.
Now assume $V_\lambda^H \neq 0$ and let $h \in V_\lambda^H - \{0\}$. Fix an highest weight vector $v_\lambda \in V_\lambda$ and let $U \subset V_\lambda$ be the unique T-stable complement to v_λ. Clearly U is B^- stable and $B^-H \subset G$ is dense in G. Then assume $h \in U$ but an the other hand B^-Hh spans V_λ a contradiction. Hence

$$h = av_\lambda + u \quad , \quad a \in \mathbf{C} - \{0\},\ u \in U$$

Since $T_o \subset H$ and h is H invariant this implies $\lambda|T_o = id$ hence λ is special.

1.6. If λ is any integral dominant weight and V_λ the corresponding irreducible representation of G with highest weight λ, we define V_λ^σ to be the space V_λ with G action twisted by σ (i.e. we set $g \circ v$ in V_λ^σ to be $\sigma(g)v$, in V_λ).

LEMMA. If λ is a special weight then V_λ^σ is isomorphic to V_λ^*.

PROOF. V_λ^* can be characterized as the irreducible representation of G having $-\lambda$ as lowest weight. Now let $v_\lambda \in V_\lambda$ be a vector of weight λ, let P be the parabolic subgroup of G fixing the line through v_λ. P is generated by the Borel subgroup B and the root subgroups relative to the negative roots $-\alpha$ for which $\langle \alpha,\lambda \rangle = 0$. Thus the parabolic subgroup P^σ, transformed of P via σ, contains the root subgroups relative to the roots $\pm\beta_i$ and also to the roots α^σ, $\alpha \in \phi_1^+$. Now $\sigma(\phi_1^+) = \phi_1^-$ hence P^σ contains the opposite Borel subgroup B^-. Clearly $v_\lambda \in V_\lambda^\sigma$ is stabilized by P^σ hence v_λ is a minimal weight vector and its weight is $-\lambda$. This proves the claim.

1.7. We have just seen that, if λ is an integral dominant special weight V_λ is isomorphic, in a σ-linear way, to V_λ^*. Under this isomorphism the highest weight vector v_λ is mapped into a lowest weight vector in V_λ^*. We normalize the mapping as follows: In V_λ the line $\mathbb{C}v_\lambda$ has a unique T-stable complement $\bar{V}_\lambda$ we define $v^\lambda \in V_\lambda^*$ by: $\langle v^\lambda, v_\lambda \rangle = 1$, $\langle v^\lambda, \bar{V}_\lambda \rangle = 0$. v^λ is easily seen to be a lowest weight vector in V_λ^*. We thus define $h\colon V_\lambda^* \to V_\lambda$ to be the (unique) σ-linear isomorphism such that $h(v^\lambda) = v_\lambda$.

REMARK. If $V = \oplus V_{\lambda_i}$ is a G-module, the action of G on $\mathbb{P}(V)$ factors through $\bar{G}$ if and only if the center of G acts on each V_{λ_i} with the same character. This applies in particular when V is a tensor product of irreducible G-modules.

We now analyze the stabilizer in G, $\tilde{H}$; of the line generated by h.

LEMMA. i) $\tilde{H}$ equals the normalizer of H.

ii) We have an exact sequence $H \hookrightarrow \tilde{H} \twoheadrightarrow C$, where C is the subgroup of the center of G formed by the elements expressible as $g\sigma(g^{-1})$ for some $g \in G$.

iii) The stabilizer of the line generated by h in $\bar{G}$ is the subgroup fixed by the order two automorphism induced by σ on $\bar{G}$.

PROOF. Assume ${}^g h = \alpha h$, α a scalar. Since h is σ linear, ${}^g h = ghg^{-1} = g\sigma(g^{-1})h$. Therefore $g\sigma(g^{-1})$ acts on V_λ as a scalar. Since V_λ is irreducible this implies $g\sigma(g^{-1})$ lies in the center of G. Conversely if $g\sigma(g^{-1})$ lies in the center of G, $g \in \tilde{H}$. We claim $g \in N(H)$. In fact putting $\zeta = g\sigma(g^{-1})$ we get for each $u \in H$

$$\sigma(g^{-1}ug) = \sigma(g^{-1})u\sigma(g) = \sigma(g^{-1})\zeta^{-1}u\zeta\sigma(g) = g^{-1}ug.$$

Now assume $g \in N(H)$. To see that $g \in \tilde{H}$ it is sufficient to show that $g\sigma(g^{-1})$ lies in the center of G or equivalently that it acts trivially on $\underline{g}$ = Lie G via the adjoint representation. Decompose $\underline{g} = \underline{h} \oplus \underline{g}_1$. And

consider the subgroup K in Aut($\underline{g}$) generated by adN(H) and σ. Since adN(H) is reductive and has at most index 2 in K(N(H) is clearly σ stable) also K is reductive. We claim that both $\underline{h}$ and $\underline{g}_1$ are K stable. In fact $\underline{h}$ is clearly K stable and the reductivity of K implies that it has a K-stable complement in $\underline{g}$, but the unique σ stable complement of $\underline{h}$ is $\underline{g}_1$ so $\underline{g}_1$ is also K stable.
Now notice that since $g \in N(H)$, for each $u \in H$

$$g^{-1}ug = \sigma(g^{-1})u\sigma(g)$$

so that $g\sigma(g^{-1})$ commutes with H and acts trivially on $\underline{h}$. On the other-hand, if $x \in \underline{g}_1$, we have $\mathrm{ad}g^{-1}(x) \in \underline{g}_1$, since $\underline{g}_1$ is K stable, so

$$-\mathrm{ad}g^{-1}(x) = \sigma(\mathrm{ad}g^{-1}(x)) = -\mathrm{ad}\sigma(g^{-1})(x)$$

and hence $\mathrm{ad}g\sigma(g^{-1})(x) = x$ so $g\sigma(g^{-1})$ acts trivially also on $\underline{g}_1$, and so on $\underline{g}$. This proves i).
ii) is clear from the above.
To see iii) notice that the subgroup fixing the line generated by h in $\bar{G}$ is the image in $\bar{G}$ of $\tilde{H}$. Hence if we denote by σ' the automorphism induced by σ on $\bar{G}$ it consists of the elements such that $g\sigma'(g^{-1}) = id$ which are the elements fixed by σ'.

REMARKS. a) H has finite index in $\tilde{H}$..
b) $\tilde{H}$ is the largest subgroup of G with Lie$\tilde{H} = \underline{h}$.

PROOF. a) follows from part ii) of the previous lemma and b) from the fact that H is connected (cf. [28]).

We complete v_λ to a basis $\{v_\lambda, v_1, v_2, \ldots, v_m\}$ of weight vectors and consider the dual basis $\{v^\lambda, v^1, v^2, \ldots, v^m\}$ in V_λ^*. We have $h(v^\lambda) = v_\lambda$ and, if χ_i is the weight of v_i we have $-\chi_i$ as weight of v^i and so $-\chi_i^\sigma$ as weight of $w_i = h(v^i)$. If we identify $\hom(V_\lambda^*, V_\lambda)$ with $V_\lambda \otimes V_\lambda$ we see that h is identified with the tensor

$$h = v_\lambda \otimes v_\lambda + \sum_{i=1}^{m} w_i \otimes v_i.$$

$v_\lambda \otimes v_\lambda$ has weight 2λ while $w_i \otimes v_i$ has weight $\chi_i - \chi_i^\sigma$.

The fact that h is σ-linear implies in particular that it is an H isomorphism. This in turn means that $\bar{h}$ is fixed under H.
Recall that $v_\lambda \otimes v_\lambda$ generates in $V_\lambda \otimes V_\lambda$ the irreducible module $V_{2\lambda}$.
Now order $\alpha_1, \ldots, \alpha_j$ so that $\alpha_s - \alpha_s^\sigma$ are mutually distinct for $s \leq \ell$ (and of course by 1.4 if $j > \ell$, for each $i > \ell$ there is an index $s \leq \ell$ such that $\alpha_s - \alpha_s^\sigma = \alpha_i - \alpha_i^\sigma$). Call $\bar{\alpha}_s = \frac{1}{2}(\alpha_s - \alpha_s^\sigma)$ $s \leq \ell$ the restricted simple roots.

PROPOSITION. i) If λ is a special weight then $V_{2\lambda}$ contains a non zero element h' fixed under H.
ii) h' is unique up to scalar multiples and can be normalized to be

$$h' = v_{2\lambda} + \sum z_i$$

with $v_{2\lambda}$ a highest weight vector of $V_{2\lambda}$ and the z_i's weight vectors having distinct weights whose weight is of the form $2(\lambda - \sum_{s=1}^{\ell} n_s\bar{\alpha}_s)$, n_i non negative integers.
iii) if λ is a regular special weight then we can assume that the vectors $z_1,\dots,z_\ell$ have weight $2(\lambda - \bar{\alpha}_1),\dots,2(\lambda - \bar{\alpha}_\ell)$.

PROOF. If we put h' equal to the image of $\bar{h}$ under the unique G-equivariant projection $V_\lambda \otimes V_\lambda \to V_{2\lambda}$, i) ii) follow from the expression of $\bar{h}$ as a linear combination of weight vectors given above. To see iii) assume λ (and hence 2λ) is a regular special weight. Since h' is fixed under H, xh' = 0 for any $x \in \underline{h} = \mathrm{Lie}H$. In particular if we let $\bar{\alpha}_s$ be a simple restricted root and $\alpha_s \in \Gamma_1$ be such that $\bar{\alpha}_s = \frac{1}{2}(\alpha_s - \alpha_s^\sigma)$ we have (cf. 1.3)

$$(x_{-\alpha_s} + \sigma(x_{-\alpha_s}))h' = 0, \quad x_{-\alpha_s} \in g_{-\alpha_s}.$$

But

$$(x_{-\alpha_s} + \sigma(x_{-\alpha_s}))v_{2\lambda} = x_{-\alpha_s}v_{2\lambda}$$

since $\sigma(x_{-\alpha_s}) \in g_{-\alpha_s^\sigma}$ and $-\alpha_s^\sigma \in \Phi_1^+$. Also by the regularity of 2λ $x_{-\alpha_s}v_\lambda$ is a non zero weight vector of weight $2\lambda - \alpha_s$. It follows that for some z_i, $\sigma(x_{-\alpha_s})z_i = -x_{-\alpha_s}v_{2\lambda}$ so that z_i has weight $2(\lambda - \bar{\alpha}_s)$ proving the claim.

The analysis just performed does not exclude that V_λ itself may contain a non zero H-fixed vector h_λ. In this case we have seen that we can normalize h_λ : $h_\lambda = v_\lambda + \sum u_i^1$, u_i^1 lower weight vectors. It follows that $h_\lambda \otimes h_\lambda$ must project to h in $V_{2\lambda}$ (by uniqueness of h).

Now the dominant λ's for which $\dim V_\lambda^H = 1$ have been determined completely [9],[24], the result is as follows: Let us indicate Λ^1 such set.

Consider the Killing form restricted to $\underline{t}_1$ and thus to $\underline{t}_1^*$. We look at the restriction of Φ_1 to $\underline{t}_1$, if $\alpha \in \Phi$, let us indicate $\bar{\alpha}$ the restriction of α to $\underline{t}_1$.

If $\mu \in \underline{t}_1^*$ let us indicate by $\tilde{\mu}$ its extension to $\underline{t}$ by setting it 0 to $\underline{t}_0$.
Then the theorem in [9] is:
Consider the set of $\mu \in \underline{t}_1^*$ such that

$\frac{(\mu,\bar{\alpha})}{(\bar{\alpha},\bar{\alpha})}$ is a positive integer for all $\alpha \in \phi$

Then the set of weights $\tilde{\mu}$ of $\underline{t}$ so obtained is exactly the set Λ^1 of λ for which dim $V_\lambda^H = 1$. One can understand this theorem in a more precise way. If $\alpha \in \phi$, then $\bar{\alpha}$ is exactly $\frac{1}{2}(\alpha - \alpha^\sigma)$, and $(\bar{\alpha};\bar{\alpha}) = (\tilde{\bar{\alpha}},\tilde{\bar{\alpha}})$. Now also a weight ω is of the form $\tilde{\mu}$ if and only if $\omega = \frac{1}{2}(\omega - \omega^\sigma)$. For such weights of course $(\omega,\beta_j) = 0$. Thus we see immediately that Λ^1 is contained in the positive lattice generated by the weights ω_i if $\sigma(i) = i$ and $\omega_i - \omega_{\tilde{\sigma}(i)}$ if $\tilde{\sigma}(i) \neq i$.

To understand exactly the nature of Λ^1 we must see if

$$\frac{(\omega_i,\bar{\alpha})}{(\bar{\alpha},\bar{\alpha})} \qquad \left(\text{resp. } \frac{(\omega_i - \omega_{\tilde{\sigma}(i)},\bar{\alpha})}{(\bar{\alpha},\bar{\alpha})}\right)$$

is an integer.

Since in any case for such special weights λ we have $2\lambda \in \Lambda^1$ one knows at least that these numbers are half integers. It follows in any case that Λ^1 is the positive lattice generated by the previous weights or their doubles. i.e.

$$\Lambda^1 = \sum_{i=1}^{\ell} n_i\mu_i, \quad n_i \geq 0 \quad \text{and} \quad \mu_i = \omega_i \quad \text{or}$$

$2\,\omega_i$ (resp. $\omega_i - \omega_{\tilde{\sigma}(i)}$ or $2(\omega_i - \omega_{\tilde{\sigma}(i)})$). Recall that $\ell = \mathrm{rk}\,\Lambda^1$ is also the rank of the symmetric space.

2. THE BASIC CONSTRUCTION

2.1. We consider now a regular special weight λ and all the objects of the previous paragraph V_λ, $h' \in V_{2\lambda}$. Let now $\mathbb{P}_{2\lambda} = \mathbb{P}(V_{2\lambda})$ be the projective space of lines in $V_{2\lambda}$ and $\tilde{h} \in \mathbb{P}_{2\lambda}$ be the class of h'. The basic object of our nalysis is the orbit $G \cdot \tilde{h}$ of $\tilde{h}$ in $\mathbb{P}_{2\lambda}$ and its closure $\bar{X} = G \cdot \tilde{h}$. By construction $\bar{X}$ is a G-equivariant compactification of the homogeneous space $G \cdot \tilde{h}$, furthermore the stabilizer $\tilde{H}$ of $\tilde{h}$ is a group containing the subgroup H.

We will analyze in detail $\tilde{H}$ and in particular will see that H has finite index in $\tilde{H}$. For the moment we concentrate our attention to $\bar{X}$. Since $\bar{X}$ is closed in $\mathbb{P}_{2\lambda}$ and G stable it contains the unique closed orbit of G acting on $\mathbb{P}_{2\lambda}$, i.e. the orbit of the highest weight vector $v_\lambda \otimes v_\lambda$. Now the following general lemma is of trivial verification:

LEMMA: If X is a G variety with a unique closed orbit Y and V is an open set in X with $Y \cap V \neq \phi$ then $X = \bigcup_{g\in G} gV$.

The use of this lemma for us is in the fact that it allows us to study the singularities of X locally in V.

2.2. Let λ be a regular special weight. Consider a G module $W \simeq V_{2\lambda} \oplus \sum V_{\mu_i}$ with $\mu_i = 2\lambda - \sum n_i 2\bar{\alpha}_i$ some $n_i > 0$. Let $h \in \underset{\sim}{V}$ be an H invariant with component h' in $V_{2\lambda}$. Decompose $V_{2\lambda} = \mathbf{C}v_{2\lambda} \oplus \tilde{V}_{2\lambda}$ in a T stable way and consider the open affine set $A = v_{2\lambda} \oplus \tilde{V}_{2\lambda} \oplus \sum v_{\mu_i} \subseteq \mathbb{P}(W)$. Notice that $h \in A$ and A is B^- stable.

LEMMA: The closure in A of the T^1 orbit T^1h is isomorphic to ℓ dimensional affine space $\mathbf{A}^\ell$. The natural morphism $T^1 \to T^1h \hookrightarrow \mathbf{A}^\ell$ has coordinates $t \to (t^{-2\bar{\alpha}_1}, t^{-2\bar{\alpha}_2}, \ldots, t^{-2\bar{\alpha}_\ell})$. T^1h is identified with the open set of $\mathbf{A}^\ell$ where all coordinates are non zero.

PROOF: By prop. 1.7 we can write $h = v_{2\lambda} + \sum_i z_i'$ with z_i' weight vectors of weights $\chi_i = 2\lambda - \sum m_j^{(i)} 2\bar{\alpha}_j$ (some $m_j > 0$) and $z_1', \ldots, z_\ell'$ of weights $2\lambda - 2\bar{\alpha}_1, \ldots, 2\lambda - 2\bar{\alpha}_\ell$. Let us apply an element $t \in T'$ to h we get $th = t^{2\lambda} v_{2\lambda} + \sum t^{\chi_i} z_i'$ which, in affine coordinates, is

$$v_{2\lambda} + \sum t^{\chi_i - 2\lambda} z_i'.$$

From the previous formula $\chi_i - 2\lambda = \sum_j m_j^{(i)} (-2\bar{\alpha}_j)$, this means that the coordinates of th are monomials in the first ℓ coordinates.

This means that T^1 maps to a closed subvariety of A, isomorphic to affine space $\mathbf{A}^\ell$, via the coordinates $(t^{-2\bar{\alpha}_1}, \ldots, t^{-2\bar{\alpha}_\ell})$. Since the restricted simple roots are linearly independent the orbit T^1h is the open dense subset of $\mathbf{A}^\ell$ consisting of the elements with non zero coordinates.

REMARK. The stabilizer of h in T^1 is the finite subgroup of the elements $t \in T^1$ with $t^{2\bar{\alpha}_i} = 1$.

2.3. Let us go back to $\underline{\bar{X}} \subseteq P_{2\lambda}$. Consider the open affine set $A = v_{2\lambda} \oplus \tilde{V}_{2\lambda} \subseteq P_{2\lambda}$ and set $V = A \cap \underline{\bar{X}}$. Remark that V is B^- stable, it contains $\tilde{h}$ and so also $\mathbf{A}^\ell$, the closure of $T^1\tilde{h}$ in A, hence $v_{2\lambda} \in V$ and therefore V has a non empty intersection with the unique closed orbit or G in $P_{2\lambda}$.

Let U be the unipotent group generated by the root subgroups X_α, $\alpha \in \phi_1^-$. Since U acts on V we have a well defined map $\phi: U \times \mathbf{A}^\ell \to V$ by the formula $\phi(u,x) = u \cdot x$.

PROPOSITION: $\phi : U \times \mathbf{A}^{\ell} \to V$ is an isomorphism.

PROOF. We first will construct a map $\psi: V \to U$ such that $\psi\phi(u,x) = u$, and prove that Im ϕ is dense in V. From this the claim follows; in fact consider the map $\zeta: V \to V$ given by $\zeta(v) = \psi(v)^{-1}v$, clearly $\zeta\phi(u,x) = x$ hence ζ maps V in $\mathbf{A}^{\ell}$ and setting $\phi': V \to U \times \mathbf{A}^{\ell}$ by $\phi'(v) = (\psi(v),\zeta(v))$ we have $\phi'\cdot\phi = 1_{U\times\mathbf{A}^{\ell}}$. Since $\phi(U \times \mathbf{A}^{\ell})$ is dense in V and $\phi\cdot\phi'$ is the identity we also have $\phi\cdot\phi' = 1_V$.

2.4. From now on we make the necessary steps for the construction of ψ.

Since 2λ is special we have, by our considerations of 1.6, that $V_{2\lambda}$ is isomorphic to $V^*_{2\lambda}$ in a σ-linear way. This isomorphism defines a non degenerate bilinear form $\langle\,,\rangle$ on $V_{2\lambda}$ which is symmetric and satisfies the following properties:

$$\langle gu,v\rangle = \langle u,\sigma(g^{-1})v\rangle \quad \text{for each } g \in G,\ u,v \in V_{2\lambda}$$

$$\langle xu,v\rangle = -\langle u,\sigma(x)v\rangle \quad \text{for each } x \in \underline{g},\ u,v \in V_{2\lambda}$$

Remark that the tangent space τ in $v_{2\lambda}$ to the orbit $U\cdot v_{2\lambda}$ has as basis the elements $x_{-\alpha}v_{2\lambda}$, $\alpha \in \phi_1^+$ (since the opposite unipotent group of U is the unipotent radical of the parabolic subgroup P stabilizing the line through $v_{2\lambda}$). Let τ^o be the subspace generated by τ and $v_{2\lambda}$.

LEMMA: i) The form $\langle\,,\rangle$ restricted to τ^o is non degenerate.
ii) τ^o is stable under P.
iii) The orthogonal $\tau^{o\perp}$ (relative to the given form) is stable under $\sigma(P)$.

PROOF: i) First of all remark that if $v_1,v_2 \in V_{2\lambda}$ are weight vectors of weights χ_1,χ_2 respectively and $\langle v_1,v_2\rangle \neq 0$ we have, for $t \in T$, $t^{\chi_1}\langle v_1,v_2\rangle = \langle tv_1,v_2\rangle = \langle v_1,\sigma(t^{-1})v_2\rangle = t^{-\chi_2^{\sigma}}\langle v_1,v_2\rangle$ and so $\chi_1 = -\chi_2^{\sigma}$. This implies that $v_{2\lambda}$ is orthogonal to $\tilde{V}_{2\lambda}$ and $\langle v_{2\lambda},v_{2\lambda}\rangle \neq 0$.
It remains to verify that on τ the form is non degenerate. Using our previous remark $\langle x_{-\alpha}v_{2\lambda},x_{-\beta}v_{2\lambda}\rangle = 0$ unless $\beta = -\alpha^{\sigma}$. In this case $\langle x_{-\beta}\sigma v_{2\lambda},x_{-\beta}v_{2\lambda}\rangle = -c\langle v_{2\lambda},x_{\beta}x_{-\beta}v_{2\lambda}\rangle$ $(c \neq 0)$ and $\langle v_{2\lambda},x_{\beta}x_{-\beta}v_{2\lambda}\rangle =$ $= \langle v_{2\lambda},[x_{\beta},x_{-\beta}]v_{2\lambda}\rangle$ since $x_{\beta}v_{2\lambda} = 0$ this is $(2\lambda,\beta)\langle v_{2\lambda},v_{2\lambda}\rangle \neq 0$. Since the map $\alpha \to -\alpha^{\sigma}$ is an involution of ϕ_1^+ the first claim follows.
ii) It is sufficient to show that τ^o is stable under the action of the Lie algebra of P. Since τ^o is stable under the torus T it is enough to show the stability of τ^o with respect to the elements x_{α} with $\alpha \in \phi_0 \cup \phi_1^+$. Now $x_{\alpha}x_{-\beta}v_{2\lambda} = [x_{\alpha},x_{-\beta}]v_{2\lambda} + x_{-\beta}x_{\alpha}v_{2\lambda}$, if $\alpha \in \phi_0 \cup \phi_1^+$ we have $x_{\alpha}v_{2\lambda} = 0$.
iii) This is clear from the properties of the form.

2.5.

LEMMA. $\mathbf{A}^{\ell} \subseteq v_{2\lambda} + \tau^{o\perp}$.

PROOF. We must show that, if $h' = v_{2\lambda} + \sum_i z_i$, each $z_i \in \tau^{o\perp}$. The weight of z_i is $\chi_i = 2\lambda - \sum n_j^{(i)} 2\bar{\alpha}_j$ so the only case to verify is when $-\sum n_j^{(i)} 2\bar{\alpha}_j = -\beta$ for some $\beta \in \phi_1^+$. Suppose this happens for z_{i_o}, since h' is H stable we have $(x_\beta + \sigma(x_\beta))h' = 0$; but $(x_\beta + \sigma(x_\beta))h' = x_\beta z_{i_o} +$ terms of weight different from 2λ, thus $x_\beta z_{i_o} = 0$. By the same weight considerations the only possible non zero scalar product between z_{i_o} and the elements of the basis of τ^o is the one with $x_{-\beta} v_{2\lambda}$, for this we have $\langle x_{-\beta} v_{2\lambda}, z_{i_o} \rangle = -\langle v_{2\lambda}, \sigma(x_{-\beta}) z_{i_o} \rangle = 0$, $(\sigma(x_{-\beta}) = c x_\beta$ some c).

2.6. Now we consider the projection π of $V_{2\lambda}$ onto $V_{2\lambda}/\tau^{o\perp}$, since $U \subset \sigma(P)$ we have a U action on $V_{2\lambda}/\tau^{o\perp}$ and the projection is equivariant. Let $K = \pi(v_{2\lambda} + \tilde{V}_{2\lambda})$, K is an affine hyperplane in $V_{2\lambda}/\tau^{o\perp}$ and it is U stable.

LEMMA. The map $j: U \to K$ defined by $j(u) = \pi(uv_{2\lambda})$ is a U equivariant isomorphism.

PROOF. From 2.4 we know that τ is the tangent space of $Uv_{2\lambda}$ in $v_{2\lambda}$. This implies that j is smooth at 1. Since j is U equivariant it is everywhere smooth. Now U has no finite subgroups and $\dim U = \dim K$ so j is an open immersion. It is a well known fact that an open immersion j of affine space $\mathbf{A}^n$ into another affine space $\bar{\mathbf{A}}^n$ of the same dimension is necessarily an isomorphism, we recall the proof: It the complement of $j(A^n)$ is non emty it is a divisor which has an equation f, this is a unit a A^n and hence a constant, giving a contradiction.
We can now construct ψ as required in 2.3, setting $\psi(v) = j^{-1}(\pi(v))$ for any $v \in V$, the fact that $\psi\phi(u,x) = u$ follows from the U equivariance of π and j and lemma 2.5.

2.7.

LEMMA. The image of ϕ is dense in V.

PROOF. The tangent space to $\mathbf{A}^{\ell}$ in $v_{2\lambda}$ is orthogonal to τ (cf. 2.5). This implies that the differential of ϕ in the point (1,0) is injective and so $\dim(\overline{\mathrm{Im}\phi}) = \dim(U \times \mathbf{A}^{\ell})$; now $\dim V = \dim \underline{X} \leq \dim G/H = \dim(U \times \mathbf{A}^{\ell})$. Since V is irreducible we get that $V = \overline{\mathrm{Im}\phi}$.

PROPOSITION. The stabilizer of $\tilde{h}$ is $\tilde{H}$.

PROOF. We have shown in the previous lemma that $\dim X = \dim G/H$ hence the subgroup H has finite index in the stabilizer of $\tilde{h}$. From 1.7 the proposition follows.

2.8. Using proposition 2.3 we identify V with the affine space $U \times \mathbb{A}^{\ell}$.

PROPOSITION. The intersection between the orbit $G\tilde{h}$ and $U \times \mathbb{A}^{\ell}$ is the open set where the last ℓ coordinates are non zero.

PROOF. We go back to $h \in \hom(V_{\lambda}^{*}, V_{\lambda}) \simeq V_{\lambda} \otimes V_{\lambda}$ (cf. 1.7) and proceed as in 2.1, 2.2. Let $h^{\#}$ be the class of h in $\mathbb{P}(\hom(V_{\lambda}^{*}, V_{\lambda})) = \mathbb{P}(V_{\lambda} \otimes V_{\lambda})$ and $\bar{X}^{\#} = \overline{G \cdot h^{\#}}$. Setting $V_{\lambda} \otimes V_{\lambda} = V_{2\lambda} \oplus Z$, the decomposition in G submodules, we consider the affine space $A^{\#} = v_{2\lambda} + \tilde{V}_2 \oplus Z$ and the G equivariant projection $\rho\colon \mathbb{P}(V_{\lambda} \otimes V_{\lambda}) \to \mathbb{P}(V_{2\lambda})$ from $\mathbb{P}(Z)$, ρ is defined in the open set $\mathbb{P}(V_{\lambda} \otimes V_{\lambda}) - \mathbb{P}(Z)$, hence in particular in $V^{\#} = \bar{X}^{\#} \cap A^{\#}$.

From the analysis of 2.2 the closure in $A^{\#}$ of the orbit $T^{1}h^{\#}$ projects under ρ isomorphically onto $\mathbb{A}^{\ell}$ hence the isomorphism $\phi\colon U \times \mathbb{A}^{\ell} \to V$ factors through $\phi\colon U \times \mathbb{A}^{\ell} \xrightarrow{\phi^{\#}} V^{\#} \xrightarrow{\rho} V$. We know that $\dim V^{\#} = \dim X^{\#} = \dim G/\tilde{H}$ (cf. 1.7) so $\operatorname{Im}\phi^{\#}$ is dense in $V^{\#}$ and as in 2.3 this implies that $\phi^{\#}$ is an isomorphism. We now have that the union of the translates of $V^{\#}$ under G is an open dense subset in $X^{\#}$ isomorphic, under ρ, to $\bar{X}$; since $\bar{X}$ is complete this open set must be $\bar{X}^{\#}$. We can now prove the proposition working with $V^{\#}$, $\bar{X}^{\#}$ and $Gh^{\#}$. The points in $U \times \mathbb{A}^{\ell}$ where the last ℓ coordinates are non zero are in the B^{-} orbit of $h^{\#}$ hence in $Gh^{\#}$, we show now that the remaining points cannot be in $Gh^{\#}$. In order to do this we interpret such points as maps from V_{λ}^{*} to V_{λ} and show that an element of $\mathbb{A}^{\ell}$ with a zero coordinate is not of maximal rank, this is clear from the analysis of 1.7. Since every point in $V^{\#}$ is in the U orbit of a point in $\mathbb{A}^{\ell}$ the proposition follows.

3. THE MINIMAL COMPACTIFICATION

3.1. We can now completely describe the structure of the variety $\bar{X}$.

THEOREM.

i) $\bar{X}$ is smooth.

ii) $\bar{X} - G \cdot \tilde{h}$ is a union of ℓ smooth hypersurfaces S_i which cross transversely.

iii) The G orbits of $\bar{X}$ correspond to the subsets of the indeces $1,2,\ldots,\ell$ so that the orbit closures are the intersections $S_{i_1} \cap S_{i_2} \cap \ldots \cap S_{i_k}$.

iv) The unique closed orbity $Y \simeq G/P$ is $\bigcap_{i=1}^{\ell} S_i$.

PROOF. We have seen that the complement of $G \cdot \tilde{h} \cap V$ in V is the union of ℓ hypersurfaces which are in fact coordinate hyperplanes, since $V \simeq U \times \mathbb{A}^{\ell}$ and the ℓ hypersurfaces Σ_i are given by the equations $x_i = 0$ for the last ℓ coordinates. Furthermore, the description of the torus action of T_1 on $\mathbb{A}^{\ell}$ shows that, two points in V are in the same $U \times T_1$

orbit if and only if they lie in the same set of hyperplanes Σ_i. Now we claim that the hypersurfaces S_i are just the closure of the Σ_i in $\bar{X}$. In fact, let S_i be any irreducible component of $S - G \cdot \tilde{h}$, necessarily S_i is G stable, since G is connected. Hence, $S_i \underset{\neq}{\supset} Y$ (the unique closed orbit) and $S_i \cap V$ is thus a component of $V - G \cdot \tilde{h}$. Hence, $S_i \cap V = \Sigma_i$ (up to reordering the indeces). Hence, $S_i = \bar{\Sigma}_i$ and conversely by the same argument, $\bar{\Sigma}_i$ is an irreducible component of $X - G \cdot \tilde{h}$, hence, it is G-stable.

To finish it is only necessary to remark that, since any point is G-conjugate to a point in V, the statement that two points in $\bar{X}$ are in the same orbit if and only if they are contained in the same S_i's follows from the similar statement relative to $U \times T_1$ in V.

3.2. Summarizing, we have found ℓ hypersurfaces S_i which are smooth. The orbits are just

$$O_{i_1,\dots,i_k} = S_{i_1} \cap \dots \cap S_{i_k} - \bigcup_{i \neq i_1,\dots,i_k} S_{i_1} \cap \dots \cap S_{i_k} \cap S_i$$

and $\bar{O}_{i_1,\dots,i_k} = S_{i_1} \cap \dots \cap S_{i_k}$ is smooth.
These are the only irreducible, closed G-stable subsets of $\bar{X}$. Their inclusion relations are, therefore, opposite to those of the faces of the simplex on the indeces $1,2,\dots,\ell$. The statement iv) is then clear.

3.3. We have just seen that, given a regular special weight λ we can describe the structure of the variety $\bar{X} = \overline{G\tilde{h}} \subset \mathbb{P}(V_{2\lambda})$. Assume now that V_λ itself contains a non zero H-invariant line generated by h' and consider $\bar{X}' = \overline{G \cdot \tilde{h}'} \subset \mathbb{P}(V_\lambda)$.

PROPOSITION. There is a natural G-isomorphism ψ: $\bar{X}' \to \bar{X}$.

PROOF. Let us consider the map ϕ: $V_\lambda \to V_{2\lambda}$ which is the composition of the map f: $V_\lambda \to V_\lambda \otimes V_\lambda$ defined by $f(v) = v \otimes v$ and of the G-equivariant projection π: $V_\lambda \otimes V_\lambda \to V_{2\lambda}$. Clearly ϕ is G-equivariant and we can normalize h' so that $\phi(h') = \bar{h}$. If we identify V_λ (resp. $V_{2\lambda}$) with $H^0(G/B,L_\lambda)$ (resp. $H^0(G/B,L_{2\lambda})$ (where L_μ is the line bundle relative to the dominant weight μ), we see that ϕ is the map taking a section into its square. Since G/H is irreducible, we then have that ϕ induces an embedding $\bar{\phi}$: $\mathbb{P}(V_\lambda) \to \mathbb{P}(V_{2\lambda})$ which is G-equivariant (and an isomorphism of $\mathbb{P}(V_\lambda)$ onto its image). Clearly $\bar{X}$ is contained in the image of $\bar{\phi}$ and is the image of $\bar{X}'$. Thus $\bar{\phi}$ induces the required isomorphism ψ.

3.4. We should remark that in the special case of a group G, considered as symmetric variety over $G \times G$, one can more simply describe the construction ad follows. If λ is a regular dominant weight of G and V_λ the corresponding irreducible representation, we consider End $(V_\lambda) = V_\lambda \otimes V_\lambda^*$ as $G \times G$ module. G is then thought as the orbit of the identity $1 \in \text{End}\,(V_\lambda)$ and the compactification $X = \overline{G \cdot \bar{1}}$ can thus be thought as "degenerate" projective transformation of the flag variety. We will refer to this case as the "compactification of $\bar{G}$".

4. INDEPENDENCE ON λ

4.1. A priori the construction performed in §2 depends on the regular weight λ, we want to show now a different construction of $\bar{X}$ which shows its independence on λ. Consider again the permutation $\tilde{\sigma}$ considered in 1.3. Each orbit of $\tilde{\sigma}$ consists of either one or two indices. Indexing the orbit by the indeces $\{1,\dots,\ell\}$, for each such index j we let λ be the sum of the fundamental weights (one or two) in the corresponding orbit. Thus a special weight is just a positive integral combination $\sum n_j \cdot \lambda_j$ while a regular one has the condition $n_j \neq 0$ for all j.

For each j we have $V_{\lambda_j} \simeq V_{\lambda_j}^*$ and a corresponding element $h_j \in V_{2\lambda_j}$. Consider then $\bar{h}_j \in \mathbb{P}(V_{2\lambda_j})$ and $\bar{h}' = (h_1,\dots,h_\ell) \in \Pi\, \mathbb{P}(V_{2\lambda_j})$. We claim that $\bar{X}$ is isomorphic to $\overline{G \cdot \bar{h}'} \subseteq \Pi\, \mathbb{P}(V_{2\lambda_j})$. In fact, consider $\lambda = \sum n_j \lambda_j$ and $\bigotimes_j V_{\lambda_j}^{\otimes n_j} = \mathbf{Q}$. Clearly $\mathbf{Q} = V_\lambda \oplus \mathbf{Q}'$ with $\mathbf{Q}'$ a sum of representations with lower highest weights. The element

$$\otimes h_j^{\otimes n_j} : \bigotimes_j V_{\lambda_j}^{\otimes n_j} \to \bigotimes_j V_{\lambda_j}^{*\otimes n_j}$$

and in particular it maps V_λ in V_λ^* and by the uniqueness of h it coincides with h on V_λ. Now we have clearly a mapping $\pi\, \mathbb{P}(V_{2\lambda_j}) \to \mathbb{P}(\otimes V_{2\lambda_j}^{\otimes n_j})$ sending $\bar{h}'$ to $\otimes h_j^{\otimes n_j}$ and so $\overline{G \cdot \bar{h}'}$ is identical to the closure of the orbit of $\otimes h_j^{\otimes n_j}$. Let $\bar{X}'$ be $\overline{G \cdot \otimes h_j^{\otimes n_j}} \subseteq \mathbb{P}(\bigotimes_j V_{2\lambda_j}^{\otimes n_j})$. We wish to project $\bar{X}'$ to X proving that they are isomorphic. In fact, we prove a more general statement which will be used later. Let us give a regular special weight λ and a representation W, with a line $\mathbf{C}h_W$ fixed under H, such that its T_1 weights are all of the form $\lambda - \sum n_i 2\bar{\alpha}_i$.

Suppose $h_\lambda \in V_\lambda$ is an H-invariant non zero vector and set $h = h_\lambda + h_W \in V_\lambda \oplus W$ and $\bar{X}' = \overline{G\tilde{h}} \subseteq \mathbb{P}(V_\lambda \oplus W)$. If we project $\mathbb{P}(V_\lambda \oplus W)$ to $\mathbb{P}(V_\lambda)$ from $\mathbb{P}(W)$ we have

LEMMA. The projection is defined on $\bar{X}'$ and establishes an isomorphism between $\bar{X}'$ and $\bar{X} = G\tilde{h}_\lambda$.

PROOF. We can assume $W = \oplus V_i$, each V_i irreducible and containing a H fixed line $\mathbf{C}h_i$ so that the projection $\Pi_i: W \to V_i$ with kernel $\underset{j\neq i}{\oplus} V_i$ has the property $\Pi_i(h_W) = h_i$.

By reasoning as in 3.3 we can double all weights and assume $\lambda = 2\lambda'$ and V_i has weight $2\mu_i$. In this situation we can define in $\bar{X}'$ the affine set V' as in 2.2 and carry out the same analysis verbatim due to the structure of the weights of h_W. Then we see that under the given map $\overset{o}{\bar{X}'} = \underset{g\in G}{\cup} V'^g$ in $\bar{X}'$ projects isomorphically onto $\bar{X}$. Since $\bar{X}$ is complete, it follows that $\overset{o}{\bar{X}'}$ is also complete and hence $\overset{o}{\bar{X}'} = \bar{X}'$ as desired.

5. THE STABLE SUBVARIETIES

5.1. We have seen that in $\bar{X}$ the only G stable subvarieties are of the form $W_{i_1,\dots,i_k} = S_{i_1} \cap S_{i_2} \cap \dots \cap S_{i_k}$ for a subset of the indices $1,2,\dots,\ell$. We wish now to describe geometrically such a subvariety. Let us then consider the weights λ_j, $j = 1,2,\dots,\ell$ defined in 4.1 and the two weights $\lambda_1 = \lambda_{i_1} + \lambda_{i_2} + \dots + \lambda_{i_k}$ and $\lambda_2 = \lambda_{j_1} + \dots + \lambda_{i_{\ell-k}}$ where $j_1,\dots,j_{\ell-k}$ are the complement of $i_1,i_2,\dots,i_k$ in $i,2,\dots,\ell$. We can, as before, consider $\bar{X}$ embedded in $\mathbb{P}(V_{2\lambda_1}) \times \mathbb{P}(V_{2\lambda_2}) \subseteq \mathbb{P}(V_{2\lambda_1} \otimes V_{2\lambda_2})$ and we can project $\bar{X}$ to $\mathbb{P}(V_{2\lambda_1})$. Let us call Π_1 this projection which is clearly G equivariant and maps onto the closure of the orbit $\bar{X}_1 = \overline{G \cdot \tilde{h}_{2\lambda_1}}$.

LEMMA. $\Pi_1(W_{i_1,\dots,i_k})$ equals the unique closed orbit in $\bar{X}_1$ (i.e. G/P_1, P_1 the parabolic, stabilizing the line through a highest weight vector in $V_{2\lambda_1}$).

PROOF: We may analyze the projection locally in V and in fact, since $V = U \cdot \mathbf{A}^\ell$, it is enough to study $\Pi_1(A^\ell \cap W_{i_1,\dots,i_k}) = \Pi_1(A^\ell_{i_1,\dots,i_k})$. We know that the intersection $\mathbf{A}^\ell \cap W_{i_1,\dots,i_k}$ is that part $A_{i_1,\dots,i_k}$ of $\mathbf{A}^\ell$ where the coordinates x_i (corresponding to $t^{-2\alpha_i}$) vanish, for $i = i_1,i_2,\dots,i_k$. The weights of the representation $V_{2\lambda_1}$, different from the highest weight, are of all of the form $\psi = 2\lambda_1 - \sum n_i\alpha_i - \sum_i \beta_i$ where at least one of the coordinates n_i relative to the indices i, for which $(\alpha_i,\lambda_i) \neq 0$, is non negative.

If we consider the projection of the subspace $A^\ell = R_1 = \overline{T_1\tilde{h}_{2\lambda}}$, this can be analyzed as follows. We have the orbit $T_1 \cdot \bar{h}_{2\lambda_1}$ and its closure R_1' and R_1 maps to R_1'. In coordinates we know that the T_1 weights appearing in $L_{2\lambda_1}$ are of type $2\lambda_1 - \sum n_i 2\alpha_i$ and then the corresponding mapping expresses such coordinates as $\Pi x_i^{n_i}$, but we know that some $n_i > 0$ for one the indices $i = i_1, i_2, \dots, i_k$. Thus we deduce that $\Pi_1(A^\ell \cap W_{i_1,\dots,i_k})$ is just the point $\overline{v_{2\lambda_1} \otimes v_{2\lambda_1}}$. This proves the lemma.

5.2. We have thus established a G equivariant mapping

$$\Pi_1 : W_{i_1,\dots,i_k} \to G \cdot \overline{v_{2\lambda_1} \otimes v_{2\lambda_1}}.$$

This last variety is of the form $G/P_{i_1,\dots,i_k}$ for the parabolic fixing $\overline{v_{2\lambda_1} \otimes v_{2\lambda_1}}$.

Since the map is G equivariant, it is a fibration. We want to study a typical fiber. Let us study $\Pi_1^{-1}\overline{(v_{2\lambda_1} \otimes v_{2\lambda_1})} = \bar{X}_1$.

Since Π_1 is a smooth morphism $\bar{X}_1$ is smooth and is the closure of the fiber of Π_1 restricted to the open orbit in $W_{i_1,\dots i_k}$; this is irreducible since P is connected. We start to study $\bar{X}_1$ locally always in the open set V. A point (γ, a) in $U_\Gamma \times A_{i_1,\dots i_k}$ is in the fiber $\bar{X}_1$ if and only if $\gamma \cdot \overline{v_{2\lambda_1} \otimes v_{2\lambda_1}} = \overline{v_{2\lambda_1} \otimes v_{2\lambda_1}}$, i.e. if and only if $\gamma \in P_{i_1,\dots,i_k}$. Now $U \cap P_{i_1,\dots,i_k}$ is exactly the unipotent subgroup generate by the root subgroup of the roots $-\alpha_i$ where $\alpha_i \in \Gamma_1$ and also α_i is a root of the Levi subgroup of $P_{i_1,\dots,i_k}$. The semisimple part of the Levi subgroup of $P_{i_1,\dots i_k}$ is relative to the root system generated by the roots β_j and the roots α_k's for which $(\alpha_k, \lambda_1) = 0$. Clearly such a subgroup $L_{i_1,\dots,i_k}$ is σ stable. Moreover, if we consider $A_{i_1,\dots,i_k} \subseteq \mathbb{P}(V_{2\lambda_2})$, we can analyze it as follows:

$h_{2\lambda_2} = v_{2\lambda_2} \otimes v_{2\lambda_2} + \sum z_i'$ where z_i' has T_1 weight $2\lambda_2 - \sum m_j 2\alpha_j$. We can split $h_{2\lambda_2}$ as $h_{2\lambda_2} = h'_{2\lambda_2} + a'$ where a' is the sum of all terms of weight $2\lambda_2 - \sum m_j 2\alpha_j$ with $m_j \neq 0$ for some $j \in \{i_1, i_2, \dots, i_k\}$. Consider any element $t \in T_1$ such that t commutes with the Levi subgroup $L_{i_1,\dots,i_k}$. Consider $H_{i_1,\dots,i_k} = L_{i_1,\dots,i_k} \cap H$, we have if $g \in H_{i_1,\dots,i_k}$, $t^{-1} \cdot gt = g$ and so $t\, h_{2\lambda_2} = g \cdot t \cdot h_{2\lambda_2}$. Hence,

$$h'_{2\lambda_2} + t \cdot a' = g \cdot h'_{2\lambda_2} + g \cdot t \cdot a'.$$

We deduce that $h'_{2\lambda_2} = g \cdot h'_{2\lambda_2}$ so $h'_{2\lambda_2}$ is $H_{i_1,\dots,i_k}$ invariant. Moreover, we see that $A_{i_1,\dots,i_k}$ can be considered as the closure of the action of the Torus $(T_1)_{i_1\dots i_k}$ on $\tilde{h}'_{2\lambda_2}$.

Thus, we deduce that the fibre we are studing is in fact the closure of the orbit of the semisimple part of the Levi subgroup acting on $\tilde{h}'_{2\lambda_2}$. Since it is easily verified that $(T_1)_{i_1\dots i_k}$ is a maximal anisotropic in $L_{i_1 \ldots i_k}$ and λ_2 restricted to $T \cap L_{i_1 \ldots i_k}$ is a regular special weight we can apply the general remarks and lemma 5.1, and see that X_1 is isomorphic to the minimal compactification of the corresponding symmetric algebraic variety $\bar{L}_{i_1 \ldots i_k}/\bar{H}_{i_1 \ldots i_k}$.

Thus we have proved:

THOREM. Let $\{i_1, \ldots, i_k\}$ be a subset of the indices $\{1,2,\ldots,\ell\}$ and let $S_{i_1,\ldots,i_k}$ be the corresponding stable subvariety of $\bar{X}$. Let $P_{i_1 \ldots i_k}$ be the parabolic subgroup associated to the weight $\lambda_1 = \lambda_{i_1} + \lambda_{i_2} + \ldots + \lambda_{i_k}$, then there is a G-equivariant fibration $\Pi_1 : S_{i_1,\ldots,i_k} \to G/P_{i_1,\ldots,i_k}$ with fibres isomorphic to the minimal compactification of $\bar{L}_{i_1 \ldots i_k}/\bar{H}_{i_1 \ldots i_k}$.

We should remark that in the case of the "compactification of a group $\bar{G}$", the set $\{1,\ldots,\ell\}$ can also be thought as the set of simple roots of G, for each subset the parabolic of $G \times G$ is $P \times P$ and the fiber of the $G \times G$ equivariant fibration is the "compactification of the adjoint group associated to the Levi factor of P".

5.3.

DEFINITION. $\bar{X}$ will be called simple if $\underline{g}$ = Lie G contains no proper σ-stable ideal.

It is clear that in this case either G is simple or we are in the case of a "compactification of a simple group".It also clear that in general $\bar{X}$ is the direct product of simple compactifications.

6. THE VARIETY OF LIE SUBALGEBRAS

6.1. We wish to compare our method with the one developed by Demazure in [5] and show that, in fact, his construction falls under our analysis.

The method is the following: consider the Lie algebras $\underline{g}$ and $\underline{h}$ of G, H respectively. Say dim $\underline{g} = n$, dim $\underline{h} = m$. Take for every $g \in G$ the subgroup gHg^{-1} and its Lie algebra $\mathrm{ad}(g)\underline{h}$. The stabilizer in G of the subalgebra $\underline{h}$ under the adjoint action is exactly the subgroup $\tilde{H}$ considered in 2.1, so we can identify $G/\tilde{H}$ with the orbit of $\underline{h}$ in the Grassmann variety $G_{m,n}$ of m-dimensional subspaces in the n-dimensional space $\underline{g}$.

We define a compactification $\tilde{\bar{X}}$ of $G/\tilde{H}$ by putting $\tilde{\bar{X}} = \overline{G\underline{h}} \subseteq G_{n,m}$.

We want to show that $\tilde{\bar{X}}$ coincides with our $\bar{X}$. If we use the Plücker embedding, we see that we can identify $\tilde{\bar{X}}$ with the closure of the G-orbit of the point $\mathbb{P}(\overset{m}{\Lambda}\,\underline{h})$ in $\mathbb{P}(\overset{m}{\Lambda}\,\underline{g})$. If h is a vector spanning the line $\stackrel{m}{\Lambda}\,\underline{h}$, h is H invariant and we want to study its weight structure.

From Proposition 1.3 we know that

$$\underline{h} = \underline{t}_0 \oplus \sum_{\alpha\in\Phi_0} g_\alpha \oplus \sum_{\alpha\in\Phi_1^+} \mathbb{C}(x_\alpha + \sigma(x_\alpha))$$

so if

$$\{\beta_1,\dots,\beta_r\} = \Phi_0^+, \qquad \{\alpha_1,\dots,\alpha_t\} = \Phi_1^+$$

We have

$$\stackrel{m}{\Lambda}\,\underline{h} = \stackrel{k}{\Lambda}\,\underline{t}_0 \Lambda x_{\beta_1} \Lambda \dots \Lambda x_{\beta_r} \Lambda x_{-\beta_1} \Lambda \dots \Lambda x_{-\beta_r} \Lambda (x_{\alpha_1} + \sigma(x_{\alpha_1}))$$

$$\Lambda \dots \Lambda(x_{\alpha_t} + \sigma(x_{\alpha_t})).$$

If we develop h and write it as a sum of weight vectors, we see that this sum contains a unique vector of weight $\mu = \alpha_1+\alpha_2+\dots+\alpha_t$ i.e. $\stackrel{\ell}{\Lambda} t_0 \Lambda x_{\beta_1} \Lambda\dots\Lambda x_{\beta_r} \Lambda x_{-\beta_1} \Lambda\dots\dots\dots\Lambda x_{-\beta_r} \Lambda x_{\alpha_1} \Lambda\dots\Lambda x_{\alpha_t}$ and the others have T_1 weitht of the form $\mu - 2\sum m_i\alpha_j$, $\alpha_j \in \Gamma_1$ and m_j non negative integers.

LEMMA. μ is a regular special weight.

PROOF. The fact that μ is special follows since $\mu^\sigma = -\mu$. To see that μ is regular recall that $2\rho = \beta_1 + \dots + \beta_r + \alpha_1 + \dots + \alpha_t$ and $(2\rho,\check{\alpha}_j) = 2$ while $(\beta_i,\check{\alpha}_j) \leq 0$ for each $\alpha_j \in \Gamma_1$ and $\beta_j \in \Phi_0^+$. Hence, clearly $(\mu,\check{\alpha}_j) \geq 2$ for each $\alpha_j \in \Gamma_1$.

We are now ready to deduce:

PROPOSITION. The compactification $\underline{\tilde{X}} = \overline{G\cdot\underline{h}} \subseteq G_{m,n}$ is isomorphic to $\underline{\overline{X}}$ of 2.1.

PROOF. Let $W \subset \stackrel{m}{\Lambda}\,\underline{g}$ be the minimum G-stable submodule containing $\mathbb{C}h = \stackrel{m}{\Lambda}\,\underline{h}$. Clearly for every irreducible component $V_i \subset W$ and G-equivariant projection Π_i: $W \to V_i$ we have $\Pi_i(h) \neq 0$.

In particular it follows from 1.5 that V_i has as its highest weight a special weight $\leq \mu$. Also, μ is a highest weight for W, we can now apply 4.1 and conclude the proof.

6.2. We can now easily see that the boundary points of $\tilde{X}$ are the Lie subalgebras (of groups related to the ones discussed in 6.2) as in Demazure's analysis.

In fact, to pass to the limit, up to conjugation, it is enough to do it under the action of T_1. If $t \in T_1$, we have:

$$t(\overset{m}{\Lambda}\,\underline{h}) = \overset{k}{\Lambda}\,\underline{t}_0 \wedge x_{\beta_1} \wedge \ldots\ldots \wedge x_{-\beta_r}$$

$$\wedge (x_{\alpha_1} + t^{-2\alpha_1}\sigma(x_{\alpha_1}) \wedge \ldots \wedge (x_{\alpha_t} + t^{-2\alpha_t}\sigma(x_{\alpha_t}))$$

Going to the limit $t^{-2\alpha_i} \to 0$ if $i = i_1,\ldots,i_k$ and $t^{-2\alpha_i} \to 1$ otherwise, we obtain the subalgebra spanned by

$$\underline{t}_0,\ x_{\beta_1},\ldots,x_{\beta_r},x_{-\beta_1},\ldots,x_{-\beta_r},x_{\alpha_k},\ldots,x_{\alpha_j}+\sigma(x_{\alpha_j})$$

where k runs over all the indices for which α_k is a root of the unipotent radical $U_{i_1,\ldots,i_k}$ of the parabolic $P_{i_1,\ldots,i_k}$ and j runs over the remaining indeces.

This is the Lie algebra of the following subgroup. Consider the automorphism σ induced on $P_{i_1,\ldots,i_k}/U_{i_1,\ldots,i_k}$. Consider the fixed points of σ in $P_{i_1\ldots i_k}/U_{i_1\ldots i_k}$ and the subgroup of $P_{i_1,\ldots,i_k}$ mapping onto this group of fixed points.

The Lie algebra is the one required by the previous analysis.

Remark that the projection from a G-orbit in $\bar{X}$ to the corresponding variety of parabolics is the one obtained by associating to a Lie algebra the normalizer of its unipotent radical.

7. COHOMOLOGY AND PICARD GROUP

7.1. We want now to describe a cellular decomposition of $\bar{X}$ which can be constructed, using the theory of Bialynicki-Birula [2],[26]. One of his main theorems is the following:

THEOREM. If $\bar{X}$ is a smooth projective variety with an action of a Torus T and if $\bar{X}$ has only a finite number of fixed points $\{x_1,\ldots,x_n\}$ under T, one can construct a decomposition $\bar{X} = \cup C_{x_i}$ where each C_{x_i} is an affine cell (an affine space) centered in x_i.

The decomposition depends on certain choices. In particular, for a suitable choice of a one parameter group μ: $G_m \to T$ such that $\bar{X}^{G_m} = \bar{X}^T$. Given such a choice, one decomposes the tangent space T_{x_i} of $\bar{X}$ at x_i as $T_{x_i} = T^+_{x_i} \oplus T^-_{x_i}$ (where T^+ and T^- are generated by vectors of positive respectively negative weight). Then C_{x_i} is an affine space of (complex) dimension $\dim T^+_{x_i}$.

Furthermore, in [26], be shows that the variety $\bar{X}$ is obtained by a sequence of attachments of the C_{x_i}'s and so the integral homology has, as basis, the fundamental classes of the closures of the C_{x_i}'s (in particular it is concentrated in even dimensions and has no torsion).

7.2. In order to apply 7.1 we need the following proposition due to D. Luna.

PROPOSITION. Let G be a reductive algebraic group acting on a variety with finitely many orbits. If T is a maximal Torus of G, the set of fixed points X^T is finite.

PROOF. We can clearly reduce to the case in which X is itself an orbit. In this case it is enough to show that, if $x \in X^T$, x is an isolated fixed point. We have $X = Gx$ by assumption and $T \subseteq St_x$. The tangent space of X in x can be identified in a T equivariant way with Lie G/Lie St_x which is a quotient of Lie G/Lie T over which T acts without any invariant subspaces, proving the claim.

In particular we can apply this proposition to our variety $\bar{X}$ in view of 3.1.

We should remark that in the case of a group G considered as $G \times G$ space, there are no fixed points on any non closed orbits. So the fixed points all lie in the closed orbit isomorphic to $G/B \times G/B$ and they are thus indexed by pairs of elements of the Weyl group.

7.3. Notice that, since $\bar{X}$ has a paving by affine spaces, we have Pic $(\bar{X}) \simeq H^2(\bar{X})$. We want now to compute $H^2(\bar{X})$ by computing the number of 2 dimensional cells given by 7.1.

For this we fix a Borel subgroup and the positive roots as in § 1. Since the center of G acts trivially on $\bar{X}$, we can use the action of a maximal Torus T of the adjoint group. Hence, the simple roots are a basis of $\underline{t}^*$. We can construct a generic 1-parameter subgroup $\mu\colon G_m \to T$ which has the same fixed points on $\bar{X}$ as T and in the following way:

We order lexicographically the simple roots as

$$\beta_1 > \beta_2 > \dots \beta_k > \alpha_1 > \dots > \alpha_\ell > \alpha_{\ell+1} > \dots > \alpha_h$$

where $\bar{\alpha}_i = \frac{1}{2}(\alpha_i - \alpha_i^\sigma)$ $i = 1,\dots,\ell$ are the restricted simple roots.

We can, since in our computations there are only finitely many weights involved (the set Λ of weights appearing in the tangent spaces of the fixed points), select μ in such a way that $\langle\lambda,\mu\rangle > 0, \lambda \in \Lambda$ if and only if $\lambda > 0$ in the lexicographic ordering. If $x \in X$ is a fixed point of T, we analyze the tangent space τ_x as follows: x is in an orbit $\mathcal{O}$ which fibers $\Pi\colon \mathcal{O} \to G/P$ with fiber a symmetric variety $\bar{L}/\bar{L}^\sigma$, we can assume $x \in \bar{L}/\bar{L}^\sigma$ and decompose τ_x in T stable subspaces $\tau_1 \oplus \tau_2 \oplus \tau_3$ such that τ_1 is isomorphic to the tangent space of $\Pi(x)$ in $G/\underline{P}$, τ_2 is isomorphic to the tangent space of x in $\bar{L}/\bar{L}^\sigma$ and τ_3 is isomorphic to the normal space of $\mathcal{O}$ in $\bar{X}$ at the point x. To compute dim τ^+ one needs to compute dim τ_i^+ for each i. Now dim τ_1^+ is given by the theory of Bruhat cells , we claim:

LEMMA. 2 dim $\tau_2^+ = \dim \tau_2$.

PROOF. The T-structure of τ_2 is isomorphic to the structure of the tangent space at the identity of $\bar{L}/\bar{L}^\sigma$ under the conjugate Torus $\bar{T} = x^{-1}T x$. Such tangent space is isomorphic to $\underline{\ell}/\underline{\ell}^\sigma$ with $\ell = \text{Lie } \bar{L}$, $\underline{\ell}^\sigma = \text{Lie } \bar{L}^\sigma$. Since $\bar{T} \subset \bar{L}^\sigma$, we see that in the root space decomposition of $\underline{\ell}$ under $\bar{T}$ we have Lie $\bar{T} \subseteq \underline{\ell}^\sigma$. $\underline{\ell}^\sigma$ is a sum of root subspaces, and if $\underline{\ell}_\alpha \subset \underline{\ell}^\sigma$, also $\ell_{-\alpha} \subset \underline{\ell}^\sigma$. Thus, $\underline{\ell}/\underline{\ell}^\sigma$ is a sum of root spaces $\underline{\ell}_\beta \oplus \underline{\ell}_{-\beta}$. And then, if $\underline{\ell}_\beta \subset (\underline{\ell}/\underline{\ell}^\sigma)^+$, we have $\underline{\ell}_{-\beta} \subset (\ell/\underline{\ell}^\sigma)^-$ and the lemma is proved.

7.4. For the computation of the T weights in T_3 we have a simple analysis in the case in which the fixed point x lies in the closed orbit G/P.

In this case $x = w x_o$, w in the Weyl group and we have:

LEMMA. In $w x_o$ the dimension of τ_3^+ equals the number of restricted simple roots $\bar{\alpha}_i$ such that $w\bar{\alpha}_i > 0$.

PROOF. Using the notations of §.2, $x_o \in V \cong U \times A^\ell$ and is identified with the point (1,0), $(1 \in U,\ 0 \in A^\ell)$. $G/P \cap V = U \times 0$, so the normal space at x_o is isomorphic to the space $\mathbf{A}^\ell$ with the induced T-action.

Thus the normal space to a point $w x_o$ is isomorphic to $\mathbf{A}^\ell$ with the action twisted by w^{-1}. Since the T weights on $\mathbf{A}^\ell$ are the $-2\bar{\alpha}_i$ we have that the T weights in the normal space at $w x_o$ are the elements $-2w\bar{\alpha}_i$, hence the claim.

7.5. In the computation of $H^2(X)$ we need to compute the points x such that dim $\tau_x^+ = 1$. Thus, we need in particular to analyze:

LEMMA. If G/H is a symmetric variety of dimension 2, with a fixed point under a Torus T', then Lie G = $\mathfrak{sl}(2)$, Lie H = $\mathfrak{so}(2)$ = Lie T', (up to normal factors on which the automorphism σ acts trivially).

PROOF: Let us recall the consequence of the Iwasawa decomposition 1.3.

$$\underline{g} = \underline{h} \oplus (\underline{t}_1 + \sum_{\alpha\in\Phi_1^+} \mathbf{C}\, x_\alpha); \quad \text{Thus, } 2 = \dim \underline{t}_1 + |\Phi_1^+|.$$

Since we generally have $\underline{t}_1 \neq 0$ if $G/H \neq 1$ and also $|\Phi_1^+| \neq 0$ since G is semisimple, we must have $1 = \dim \underline{t}_1 = |\Phi_1^+|$. Moreover, since we want to factor out all normal subgroups of G on which σ acts trivially, we have G simple. We wish to show that Φ_0 is empty. In fact, if there is a simple root $\beta \in \Phi_0$, since G is simple we may assume that $\beta + \alpha$ is also a root. But then either β or $\beta+\alpha \in \Phi_1^+$ and we have a contradiction. Then we see that G is of rank 1 and the remaining statements easily follow.

7.6. We are now ready for the computation of Pic (X).

THEOREM. Pic (X) $\cong Z^{\ell+r}$ where r is the number of simple roots α_i, $i = 1,\dots,\ell$ such that: there exist two distinct simple roots α, β with $\bar{\alpha}_i = \frac{1}{2}(\alpha - \alpha^\sigma) = \frac{1}{2}(\beta - \beta^\sigma)$ and either $-\alpha^\sigma \neq \beta$ or, if $-\alpha^\sigma = \beta$, $(\alpha,\beta) \neq 0$.

PROOF. Let $\bar{\alpha}_i = \frac{1}{2}(\alpha_i - \alpha_i^\sigma)$, $i = 1,\dots,\ell$ be the simple restricted roots (cf. 2.2). Suppose $x \in \bar{X}$ is a fixed point with $\dim \tau_x^+ = 1$, first of all we analyze the case in which $x \in G/P$, the unique closed orbit. In this case $\tau_x = \tau_1 + \tau_3$ and we must have either $\dim \tau_1^+ = 0$, $\dim \tau_3^+ = 1$ or $\dim \tau_1^+ = 1$, $\dim \tau_3^+ = 0$. Now x is a center of a Bruhat cell in G/P of dimension equal to $\dim \tau_1^+$ so it is either the point x_o corresponding to the 0 cell or a point $s_\alpha x_o$ with α a simple root in Φ_1^+. Thus, by Lemma 7.4 $\dim \tau_3^+$ at wx_o is the number of i such that $w\bar{\alpha}_i$ is negative. In particular we see that we can get 2 dimensional cells only centered at the points $s_\alpha x_o$ and we need to count how many $\alpha \in \Phi_1^+$ are such that $s_\alpha \bar{\alpha}_i > 0$ for all i's. Now if $\alpha \neq \alpha_i, -\alpha_i^\sigma$, we have $s_\alpha(\bar{\alpha}_i) > 0$ (since $s_\alpha(\beta) > 0$ if β is positive $\alpha \neq \beta$). Now given $\alpha \in \Phi_1^+$ if $\alpha = \alpha_i$, we have $s_\alpha(\bar{\alpha}_j) > 0$ if $j \neq i$. As for $s_\alpha(\bar{\alpha}_i)$ it depends on $-\alpha_i^\sigma$.
We have various cases:

i.) $-\alpha_i^\sigma = \alpha_i$,

ii.) $-\alpha_i^\sigma = \alpha_i + \beta$, $\beta \neq 0$ a positive combination of roots in Φ_0.

iii.) $-\alpha_i^\sigma = \alpha_j + \beta$, $j \neq i$.

In case i.) $s_\alpha(\bar{\alpha}) = -\bar{\alpha} < 0$,

In case ii.) $s_\alpha(\bar{\alpha}) = -\alpha + \frac{1}{2}(\beta - \frac{2(\alpha.\beta)}{(\alpha,\alpha)}\alpha) > 0$,

In case iii.) the same reasoning as in ii.) holds if $\beta \neq 0$,
$s_\alpha(\alpha + \alpha_j + \beta) = \beta + \alpha_j + m\alpha > 0$ (some m).
If $\beta = 0$, we have

$$s_\alpha(\alpha + \alpha_j) = -\alpha + \alpha_j - \frac{2(\alpha,\alpha_j)}{(\alpha,\alpha)}\alpha$$

Now since $\alpha_j = -\alpha^\sigma$, we must have $(\alpha,\alpha) = (\alpha_j,\alpha_j)$. Hence, the Dynkin diagram formed by α, α_j is either disconnected and $(\alpha,\alpha_j) = 0$ or is A_2 and then $\frac{2(\alpha,\alpha_j)}{(\alpha,\alpha)} = -1$ so $s_\alpha(\alpha + \alpha_j) = \alpha_j > 0$. If $(\alpha,\alpha_j) = 0$, we have $s_\alpha(\alpha + \alpha_j) = -\alpha + \alpha_j < 0$ since $\alpha = \alpha_i$ $i \leq \ell$ and $j > \ell$.

Now we have to consider the case $\alpha = -\alpha_i^\sigma \neq \alpha_i$, since α is a simple root this occurs only in the case $-\alpha_i^\sigma = \alpha_j$, $j > \ell$. The same analysis as before shows that if $(\alpha,\alpha_i) = -\frac{1}{2}$ we have $s_\alpha(\alpha + \alpha_i) = \alpha > 0$

if $(\alpha,\alpha_i) = 0$ $s_\alpha(\alpha + \alpha_i) = \alpha_i - \alpha > 0.$

It remains to analyze the case of x lying in a non closed orbit $\mathcal{O}$. By Lemmas 7.3 and 7.5 this can occur only when $\mathcal{O}$ fibers on a variety G/P' with fiber the minimal compactification of a symmetric variety isomorphic to $SL(2)/S\mathcal{O}(2)$. This is the variety of distinct unordered pairs of points in $\mathbb{P}^1$ and its minimal compactification is the space $\mathbb{P}^2$ considered as the symmetric square of $\mathbb{P}^1$. In this case we only have 2 SL(2) orbits in $\mathbb{P}^2$ and so only 2 G orbits in $\bar{\mathcal{O}}$.
Thus by 3.1 we have $\dim \bar{\mathcal{O}} = \dim G/P + 1$ and a $\mathbb{P}^1$-fibration $G/P \to G/P'$. Thus, we can identify P' with the parabolic group generated by P and the subgroup $X_{-\alpha}$ relative to a simple root $\alpha \in \Phi_1^+$ and we have $\alpha^\sigma = -\alpha$, and $\alpha = \alpha_i$ for some $1 \le i \le \ell$. As in Lemma 7.3 write $\tau_x = \tau_1 \oplus \tau_2 \oplus \tau_3$. Since T acts on τ_2 by a negative and a positive weight as we have noted above in order to have that the set of T weights appearing on τ_x contains only a positive weight, we must have that the T weights in τ_1 and τ_3 consist of negative weights. This implies that $p(x) \in G/P'$ is the unique B fix point in G/P', otherwise at least one of the weights appearing in τ_1 would be positive. Furthermore, notice that the fact that p(x) is the unique B fix point in G/P' determines x uniquely since in $p^{-1}(p(x)) = \mathbb{P}^2$ there are exactly three T fix points of which two are x_0 and $s_\alpha(x_0)$ both belonging to the closed orbit. But for such x we have that the set of weights appearing on τ_3 is
$\{-\frac{(\alpha_j - \alpha_j^\sigma) + s_\alpha(\alpha_j - \alpha_j^\sigma)}{2}\}$ for $1 \le j \le \ell$, $j \ne i$ which are all negative. This is easily seen as follows: first of all the normal bundle to $\bar{\mathcal{O}}$ in X is just the sum of the restrictions of the normal line bundles to the closures of the codimension one orbits S_j, $1 \le j \le \ell$, $j \ne i$, containing $\bar{\mathcal{O}}$. Thus we have to compute the weight of T for each such line bundle N_j. Let us fix $1 \le j \le \ell$, $j \ne i$, then the T weight of N_j in x_0 is just $-(\alpha_j,-\alpha_j^\sigma)$. Now if we let $T_\alpha \subset T$ denote ker α, we have that T_α acts trivially on $\mathbb{P}^2$ hence the T_α weight in x and x_0 are the same. Thus the given formula is correct for T_α. It remains to verify the formula on a "complement of T_α in T. This amounts to perform the computation in the maximal torus of PSL(2) which can be carried out directly.
So it follows that the action of T on τ_x has exactly one negative weight and the cell associated to x has dimension 2. Summarizing our result we have

1) If $\bar{\alpha}_i$ is such that there exists only one simple root α with $\frac{1}{2}(\alpha - \alpha^\sigma) = \bar{\alpha}_i$ and $\alpha^\sigma \neq -\alpha$ then we get one 2 cell whose center lies in the unique closed orbit G/P.
2) If α_i is as in one but $\alpha^\sigma = -\alpha$ then again we get one 2 cell but its center lies in the orbit $\mathcal{O}$ whose closure $\bar{\mathcal{O}}$ fibers with $\mathbb{P}^2$ fibers onto G/P', P' being the parabolic generated by P and $X_{-\alpha}$.
3) If $\bar{\alpha}_i$ is such that there exists two distinct simple roots α,β such that $\bar{\alpha}_i = \frac{1}{2}(\alpha - \alpha^\sigma) = \frac{1}{2}(\beta - \beta^\sigma)$, $-\alpha^\sigma = \beta$ and $(\alpha,\beta) = 0$ then we get exactly one 2 cell whose center lies in G/P.
4) If $\bar{\alpha}_i$ is such that $\bar{\alpha}_i = \frac{1}{2}(\alpha - \alpha^\sigma) = \frac{1}{2}(\beta - \beta^\sigma)$ and either $-\alpha^\sigma \neq \beta$ or $-\alpha^\sigma = \beta$ but $(\alpha,\beta) \neq 0$, then we get two 2 cells, whose both centers lie in G/P.

This is our theorem.

DEFINITION. $\underline{\bar{X}}$ will be called exceptional when rk Pic$(\underline{\bar{X}}) > \ell$.

7.7. REMARK. It is clear from the previous analysis that the main difficulty in computing explicitely the dimensions of the cells lies in the computation of τ_3^+. In the special case in which all fixed points lie in the closed orbit this is accomplished by Lemma 7.4.

In particular for the case of a group $\bar{G}$ considered as a symmetric variety over $\bar{G} \times \bar{G}$ we have the following computation for the Poincarè polynomial: $\Sigma b_i q^j$, $b_i = \dim H_i(\underline{\bar{X}},\mathbb{Z})$:

$$\Big(\sum_{w\in W} q^{2\ell(w)}\Big)\Big(\sum_{w\in W} q^{2(\ell(w)+L(w))}\Big)^{(*)}$$

($\ell(w)$ the length of w, $L(w)$ the number of simple reflections s_α with $\ell(s_\alpha w) < \ell(w)$).

8. LINE BUNDLES ON $\underline{\bar{X}}$

8.1. Let $\underline{\bar{X}}$ be as usual and let $Y = G/P \subset \underline{\bar{X}}$ be the unique closed orbit in $\underline{\bar{X}}$.

PROPOSITION. Let i^*: Pic $(\underline{\bar{X}}) \to$ Pic (Y) be the homomorphism induced by the inclusion. Then i^* is injective.

PROOF. First assume that for any simple root $\alpha \in \Phi_1^+$ we have $\alpha^\sigma = -\alpha$. Then we know that Pic $(\underline{\bar{X}}) \simeq \mathbb{Z}^\ell$, where ℓ is the number of simple roots in Φ_1^+. Furthermore, let $\omega_1,\dots,\omega_\ell$ be the fundamental weights correspond-

---

(*) We wish to thank G. Lusztig for suggesting this formula.

ing to such α's. Then we have shown how to imbed $\bar{X} \subset \prod_{i=1}^{s} \mathbb{P}(V_{2\omega_i})$. So we get a map h^*: $\mathrm{Pic}(\prod_{i=1}^{s} \mathbb{P}(V_{2\omega_i})) \to \mathrm{Pic}\,(\bar{X})$. But it is clear that i^*h^* is injective since the restriction of the tautological bundle L_i on $\mathbb{P}(V_{2\omega_i})$ to G/P gives the line bundle associated to $2\omega_i$. Since $\mathrm{rk}(\mathrm{Pic}(\prod_{i=1}^{s} \mathbb{P}(V_{2\omega_i})) = \mathrm{rk}(\mathrm{Pic}\,(X))$ our assertion follows.

Let us now suppose that there exists a simple root α such that $\alpha^{\sigma} \neq -\alpha$. Let S be the unique orbit closure associated to $\alpha - \alpha^{\sigma}$. Then it follows from the description of the dimension two cells given in 7, that each dimension 2 cell in $\bar{X}$ is already contained in S, so we prove that the map $\mathrm{Pic}\,(\bar{X}) \to \mathrm{Pic}\,(S)$ induced by inclusion is injective.

Let us now consider the map $\mathrm{Pic}\,(S) \to \mathrm{Pic}\,(Y)$ and recall that for a suitable parabolic P_o we get a fibration $S \to G/P_o$ whose fiber is the variety $\bar{X}_{\bar{L}}$ which is the minimal compactification of $\bar{L}/\bar{L}^{\sigma}$ where $\bar{L}$ is the adjoint group of the semisimple Levi factor of P_o and $\bar{L}^{\sigma}$ the fix points group of the involution induced by σ on $\bar{L}$. We thus get the diagram

$$\begin{array}{ccc} P_o/P & \hookrightarrow & \bar{X}_{\bar{L}} \\ \downarrow & & \downarrow \\ Y=G/P & \hookrightarrow & S \\ & \searrow \quad \swarrow & \\ & G/P_o & \end{array}$$

and we can identify P/P_o with the unique closed orbit in $\bar{X}_{\bar{L}}$. But notice that $\mathrm{Pic}\,(G/P) \cong \mathrm{Pic}\,(G/P_o) \oplus \mathrm{Pic}\,(P/P_o)$ and $\mathrm{Pic}\,(S) \cong \mathrm{Pic}\,(G/P_o) \oplus \mathrm{Pic}\,(\bar{X}_{\bar{L}})$. Also, by induction on the rank we can assume that the map $\mathrm{Pic}\,(\bar{X}_L) \to \mathrm{Pic}\,(P/P_o)$ induced by inclusion is injective. This clearly implies that the map $\mathrm{Pic}\,(S) \to \mathrm{Pic}\,(Y)$ is also injective.

REMARK. Notice that since we can identify Pic (Y) with the lattice spanned by the fundamental weights relative to the simple roots in Φ_1^+, our proposition implies that we can also identify Pic (X) with a sublattice of such a lattice, call it Γ. Notice also that since for each dominant special weight λ with the property that $\dfrac{2(\lambda,\alpha-\alpha^{\sigma})}{(\alpha-\alpha^{\sigma},\alpha-\alpha^{\sigma})} \in \mathbb{Z}^+$ for every simple root $\alpha \in \Phi_1^+$ we have constructed a map $\Pi\colon X \to \mathbb{P}(V_{\lambda})$ we clearly have that Γ contains the lattice spanned by such weights. In particular, this lattice contains the double of the lattice of special weights $\alpha - \alpha^{\sigma} \in \Gamma$ for each simple root $\alpha \in \Phi_1^+$.

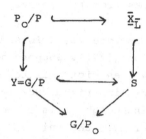

We wish to collect some of the information gotten up to now for future use.

We have the weights μ_i introduced in 1.7 and a natural embedding

$$\bar{X} \to \Pi\mathbb{P}(V_{\mu_i})$$

The mapping of the closed orbit $Y \to \Pi\mathbb{P}(V_{\mu_i})$ so induced is the canonical one obtained by the diagonal morphism. We compose this with the natural projection $G/B \to Y$.

The ample generator of Pic $(\mathbb{P}(V_{\mu_i}))$ is mapped by the composed homomorphism to the element L_{μ_i} of Pic(G/B) corresponding to the weight μ_i (notice that under this convention $H^o(G/B,L_{\mu_i}) \cong V^*_{\mu_i}$ as a G-module).

If J is a subset of $\{1,\ldots,\ell\}$ and S_J denotes the corresponding orbit closure, the composition $S_J \to \bar{X} \to \Pi\mathbb{P}(V_{\mu_i}) \xrightarrow{P_J} \prod_{i\in J}\mathbb{P}(V_{\mu_i})$ factors through the canonical fibration $S_J \to G/P_J$ and the canonical inclusion $G/P_J \hookrightarrow \prod_{i\in J}\mathbb{P}(V_{\mu_i})$. Therefore in particular the line bundle corresponding to μ_i restricted to S_J comes from the corresponding line bundle in G/P_J.

Finally since Pic$(\bar{X})$ is discrete and G is simply connected any $L \in$ Pic $(\bar{X})$ has a G linearization ([27]). Suppose now $L \in$ Pic $(\bar{X})$ is a G linearized line bundle. If we restrict this to the closed orbit Y we have the induced bundle already linearized. Now for a linearized line bundle L_λ on Y the corresponding weight λ is the character by which the maximal torus acts on the fiber over the unique B fix point, x_o, in Y.

Recall that the cell $U \times \mathbb{A}^\ell$ in $\bar{X}$ is a B^- stable affine subspace and (1,0) is the fixed point x_o in Y previously introduced. If s is a section trivializing L_λ on $U \times \mathbb{A}^\ell$ so is $b*s$ for any $b \in B^-$. Since the only invertible functions on $U \times \mathbb{A}^\ell$ are the constants we have $b*s = \alpha s$, α a scalar. Restricting to the point x_o we have $\alpha = b^{-\lambda}$.

8.2. Notice that since any $L \in$ Pic (X) can be G linearized we have that G acts linearly on each $H^i(\bar{X},L)$.

LEMMA. Let $L \in$ Pic $(\bar{X})$ and consider $H^o(\bar{X},L)$ as a G module. Then $\dim \mathrm{Hom}_G(V,H^o(\bar{X},L)) \leq 1$ for each irreducible G-module V.

PROOF. Suppose $\mathrm{Hom}_G(V,H^o(\bar{X},L)) \neq 0$. Let μ be the highest weight of V. Let $s_1,s_2 \in H^o(\bar{X},L)$ be two non zero U invariant sections whose weight is μ. Then $\frac{s_1}{s_2}$ is a B invariant rational function on $\bar{X}$. Since B has a dense orbit in $\bar{X}$, it follows the $\frac{s_1}{s_2}$ is constant. Hence, s_1 is a multiple of s_2 and our claim follows.

Now let $V \subset \bar{X}$ be the open set described in 2 and identify V with

$U \times \mathbb{A}^\ell$. Let $\{x_i\}$ be the coordinate functions on $\mathbb{A}^\ell$. For any $t \in T$, $tx_i = t^{-(\alpha_i - \alpha_i^\sigma)} x_i$ for the corresponding simple root $\alpha_i \in \Phi_1^+$, $1 \leq i \leq \ell$.

PROPOSITION. Let V_μ be the irreducible G-module whose highest weight is μ. Let $\lambda \in \Gamma$ and $L_\lambda \in \mathrm{Pic}\,(\underline{\bar{X}})$ be the corresponding line bundle, then if

$$\mathrm{Hom}(V_\mu^*, H^0(\underline{\bar{X}}, L_\lambda)) \neq 0 \qquad \mu = \lambda - \sum t_i(\alpha_i - \alpha_i^\sigma), \quad t_i \in Z^+$$

PROOF. Let $s \in H^0(\underline{\bar{X}}, L_\lambda)$ be a section generating a B^- stable line. Then if we restrict s to V and we let s_0 be a section trivializing $L_\lambda|V$ we can write $s = s_0 f$ where f is a regular function on $V \cong U \times \mathbb{A}^\ell$. Since s is U stable f is also U stable and $f = x_1^{t_1}, \ldots, x_\ell^{t_\ell}$ so our proposition follows.

COROLLARY. There exists a unique up to a scalar G-invariant section $r_i \in H^0(\underline{\bar{X}}, L_{\alpha_i - \alpha_i^\sigma})$ whose divisor is S_i.

PROOF. Let $r_i \in H^0(\underline{\bar{X}}, \mathcal{O}(S_i))$ be the unique, up to constant, section whose divisor is S_i. Since S_i is G-stable and G is semisimple, r_i is a G-invariant section. Also since $x_i = 0$ is a local equation of S_i on V we have $r_i|V = s_o x_i$ where s_o is a section trivializing $\mathcal{O}(S_i)|V$. The weight of x_i is $\alpha_i - \alpha_i^\sigma$ so the G-invariance of r_i implies that s_o has weight $-(\alpha_i - \alpha_i^\sigma)$. Hence $\mathcal{O}(S_i) \cong L_{\alpha_i - \alpha_i^\sigma}$.

8.3. Now let $S_{\{i_1,\ldots,i_t\}} = S_{i_1} \cap \ldots \cap S_{i_t}$ for any subset $\{i_1,\ldots,i_t\} \subset \{1,\ldots,\ell\}$ be the corresponding G-stable subvariety. Let $\gamma \in \Gamma$ put $L_\gamma(i_1,\ldots,i_t) = L_\gamma|S_{i_1,\ldots,i_t}$. Let $\{j_1,\ldots,j_{\ell-t}\}$ denote the complement in $\{1,\ldots,\ell\}$ of $\{i_1,\ldots,i_t\}$.

PROPOSITION. Let $\gamma \in \Gamma$ be a dominant weight. Let $\{h_1,\ldots,h_s\} \subset \{j_1,\ldots,j_{\ell-t}\}$. Then

$$H^i(S_{\{i_1,\ldots,i_t\}}, L_{\gamma - \Sigma(\alpha_{h_i} - \alpha_{h_i}^\sigma)}(i_1,\ldots,i_t)) = 0 \quad \text{for} \quad i > 0.$$

PROOF. We perform a double decreasing induction on $\{i_1,\ldots,i_t\}$ and on $\{h_1,\ldots,h_s\}$.

If $\{i_1,\ldots,i_t\} = \{1,\ldots,\ell\}$ then $\{1,\ldots,\ell\} = G/P$ is the unique closed orbit and our proposition is part of Bott's theorem [4].

Now let $\{i_1,\ldots,i_t\}$ be arbitrary and $\{j_1,\ldots,j_{\ell-t}\} = \{h_1,\ldots,h_s\}$. Then notice that by our local description of $\bar{X}$ it follows easily that if $K(i_1,\ldots,i_t)$ denotes the canonical bundle on $S_{\{i_1,\ldots,i_t\}}$, $K(i_1,\ldots,i_t) = L_{-\mu - \sum_{m=1}^{\ell-t}(\alpha_{j_m} - \alpha_{i_m}^\sigma)}(i_1,\ldots,i_t)$ where $\mu = \sum_{\alpha \in \Phi_1^+} \alpha$.

(Notice that $\mu \in \Gamma$ (cf. 6.1)).
Thus if we put $L = L_{\gamma-\Sigma(\alpha_{j_m}-\alpha^{\sigma}_{j_m})}(i_1\ldots,i_t)$ and $K = K(i_1,\ldots,i_t)$ we have that $(K \otimes L^{-1})^{-1} = L_{\gamma+\mu}(i_1,\ldots,i_t)$ can be verified to be very ample. We postpone the proof of this assertion to the end of this section. It follows from Kodaira vanishing theorem that

$$H^i(S_{\{i_1,\ldots,i_t\}}, K \otimes L^{-1}) = 0 \quad \text{for} \quad i < \dim S_{i_1,\ldots,i_t}.$$

This implies by Serre's duality

$$H^i(S_{\{i_1,\ldots,i_t\}}, L) = 0 \quad \text{for } i > 0.$$

Now by induction we have the result proved for any $S_{\{i_1,\ldots,i_{t+1}\}}$ and for any $\{h_1,\ldots,h_{s+1}\} \subset \{j_1,\ldots,j_{\ell-t}\}$.
Corollary 8.2 implies that we have a non zero section

$$r_{i_{t+1}} \in H^0(S_{\{i_1,\ldots,i_t\}}, L_{\alpha_{i_{t+1}}-\alpha^{\sigma}_{i_{t+1}}})(i_1,\ldots,i_t))$$

and multiplication by $r_{i_{t+1}}$ yields an exact sequence.

$$0 \to L_{\gamma - \sum_{i=1}^{s}(\alpha_{h_i} - \alpha^{\sigma}_{h_i} - (\alpha_{i_{t+1}} - \alpha^{\sigma}_{i_{t+1}}))}(i_1,\ldots,i_t) \to$$

$$L_{\gamma - \sum_{i=1}^{s}(\alpha_{h_i} - \alpha^{\sigma}_{h_i})}(i_1,\ldots,i_t) \to L_{\gamma - \sum_{i=1}^{s}(\alpha_{h_i} - \alpha^{\sigma}_{h_i})}(i_1,\ldots,i_{t+1}) \to 0$$

Then we get a long exact sequence that together with an inductive hypothesis immediately proves the proposition.

THEOREM. Let $\lambda \in \Gamma$ then:
1) $H^0(\bar{X}, L_\lambda) \neq 0$ if and only if $\lambda = \gamma + \Sigma t_i(\alpha_i - \alpha^{\sigma}_i)$ for some dominant γ, $t_i \in \mathbb{Z}^+$. Assuming $H^0(X,L_\lambda) \neq 0$, if V_γ is the irreducible G-module of highest weight γ, $H^0(X,L_\lambda) = \oplus V^*_\gamma$ for all dominant γ of the form $\gamma = \lambda - \Sigma t_i(\alpha_i - \alpha^{\sigma}_i)$, $t_i \in \mathbb{Z}^+$.
2) For λ dominant $H^i(\bar{X},L_\lambda) = 0$, $i > 0$.

PROOF.
1) The only if part is just Proposition 8.2.
To prove the if part assume λ is dominant. Then we know that $H^0(G/P, L_\lambda|_{G/P})$ is the irreducible G-module V_λ whose highest weight is λ. Now consider the varieties

$$\bar{X} = S_\phi \supset S_{\{1\}} \supset S_{\{1,2\}} \supset S_{\{1,2,3\}} \supset \ldots\ldots S_{\{1,2,\ldots,\ell\}} = G/P$$

We claim that for each $\ell \geq i \geq 1$ the restriction map

$$H^{0}(S_{\{1,2,\dots,i-1\}}, L_\lambda|_{S_{\{1,2,\dots,i-1\}}}) \to H^{0}(S_{\{1,2,\dots,i\}}, L_\lambda|_{S_{\{1,2,\dots,i\}}})$$

is onto.

This follows at once from the cohomology exact sequence associated to the sequence

$$0 \to L_{\lambda-(\alpha_i-\alpha_i^\sigma)}(i,2,\dots,i-1) \to L_\lambda(1,2,\dots,i-1) \to L_\lambda(1,2,\dots,i) \to 0$$

considered above and the vanishing of

$$H^1(S_{\{1,\dots,i-1\}}, L_{\lambda-(\alpha_i-\alpha_i^\sigma)}(1,2,\dots,i-1))$$

proved in Proposition 8.3.
In particular, the restriction map

$$H^0(\bar{X}, L_\lambda) \to H^0(G/P, L_\lambda|_{G/P}) \text{ is onto.}$$

Hence, $\mathrm{Hom}_G(V^*_\lambda, H^0(\bar{X}, L_\lambda)) \neq 0$ and we can find a non zero lowest weight vector $\underline{v}_\lambda \in H^0(\bar{X}, L_\lambda)$ whose weight is $-\lambda$.

Now let $\lambda = \gamma + \sum_{i=1}^{\ell} t_i(\alpha_i - \alpha_i^\sigma)$, $t_i \in \mathbb{Z}^+$, γ dominant in Γ.
Consider the section $r_1^{t_1} \dots . r_\ell^{t_\ell} \in H^0(\bar{X}, L_{\sum_{i=1}^{\ell} t_i(\alpha_i - \alpha_i^\sigma)})$ and the section $v_{-\gamma} \in H^0(\bar{X}, L_\gamma)$.
Then the section $v_{-\gamma} r_1^{t_1} \dots r_\ell^{t_\ell}$ is clearly non zero U-invariant and its weight is $-\gamma$. So $\mathrm{Hom}_G(V_\gamma, H^0(\bar{X}, L_\lambda)) \neq 0$. This proves 1); 2) is contained in Proposition 8.3.

REMARK.

1) By a completely analogous argument we can prove that if $\lambda \in \Gamma$ then

$$\mathrm{Hom}\,(V^*_\gamma, H^0(S_{\{i_1,\dots,i_t\}}, L_\lambda|_{S_{\{i_1,\dots,i_t\}}})) \neq 0$$

if and only if

$$\lambda = \gamma + \sum_{m=1}^{\ell-t} t_m(\alpha_{j_m} - \alpha_{j_m})$$

2) Clearly we can define a filtration of $H^0(\bar{X}, L_\lambda)$ by putting for each ℓ-tuple of non negative integers $(t_1, \dots, t_\ell)$, $W(t_1, \dots, t_\ell)$ to be the subspace of sections $s \in H^0(\bar{X}, L_\lambda)$ vanishing on S_1 of order $\geq t_1, \dots,$ on S_ℓ of order $\geq t_\ell$. Then we can restate our theorem as follows:

$$W_\lambda(t_1,\dots,t_\ell) \Big/ \sum_{(\bar t_1,\dots,\bar t_\ell) > (t_1,\dots,t_\ell)} W_\lambda(\bar t_1,\dots,\bar t_\ell) \cong \begin{cases} 0 & \text{if } \lambda-\sum t_i(\alpha_i-\sigma(\alpha_i)) \text{ is not dominant} \\ V^*_{\lambda-\sum t_i(\alpha_i-\sigma(\alpha_i))} & \text{otherwise} \end{cases}$$

(Here $(\bar t_1,\dots,\bar t_\ell) \geq (t_1,\dots,t_\ell)$ means $\bar t_1 \geq t_1,\dots,\bar t_\ell \geq t_\ell$).

8.4. In order to complete the proof of 8.3 we have to discuss the ampleness of $L_{\gamma+\mu}(i_1\dots i_t)$ which has been used there.

We start with a general easy fact. Let ω,ω' denote two distinct fundamental weigths V_ω, $V_{\omega'}$, $V_{\omega+\omega'}$ the irreducible representations of highest weight $\omega,\omega',\omega+\omega'$.

We have a canonical G equivariant projection $p\colon V_\omega \otimes V_{\omega'} \to V_{\omega+\omega'}$ and we denote by $\bar p$ the induced projection $\mathbb{P}(V_\omega \otimes V_{\omega'}) \to \mathbb{P}(V_{\omega+\omega'})$ of projective spaces: Remark that $\mathbb{P}(V_\omega) \times \mathbb{P}(V_{\omega'})$ is embedded in $\mathbb{P}(V_\omega \otimes V_{\omega'})$ via the Segre map.

LEMMA. The map $\bar p$ restricted to $P(V_\omega) \times P(V_{\omega'})$ is a regular embedding.

PROOF. We consider the irreducible representations of G as sections of line bundles an G/B so that the map p corresponds to the usual multiplication. Since G/B is irreducible the product of 2 non zero sections is always non zero. Now if $s,s' \in V_\omega$, $t,t' \in V_{\omega'}$ and $st = s't'$ we claim that $s' = cs$, $t' = c^{-1}t$, c a scalar. In fact since ω,ω' are fundamental the divisors of s,s',t,t' are all irreducible since ω,ω' are independent in Pic (G/B) the divisor of s cannot equal the divisor of t' and so we have divs = divs' and the claim.

This proves that $\bar p$ is injective when restricted to $P(V_\omega) \times P(V_{\omega'})$. To see that the map is also smooth one can use the same fact in local affine coordinates.

We are now ready to prove:

PROPOSITION. For any $\gamma \in \Gamma$ dominant the line bundle $L_{\gamma+\mu}$ is ample on $\bar X$ hence also on $S_{\{i_1,\dots,i_t\}}$ for any choice of $i_1,\dots,i_t$.

PROOF. We distinguish 2 cases. If γ is special, since μ is a regular special weight so is $\mu+\gamma$ hence by 3.1 and 4.1 we have that $L_{2(\mu+\gamma)}$ is very ample on X.

Assume γ not special. This can happen only if we are in the exeptional case i.e. if the $\mathrm{rkPic}(\bar X) > \ell$ since if a multiple of a weight γ is special so is γ and $\mathrm{Pic}(\bar X)$contains the double of the lattice of special weights.

First of all we can clearly reduce to the case is which X is simple (cf. 5.3).

In the group case $\underline{\bar{X}} = \overline{G \times G/G}$ we have rk Pic($\underline{\bar{X}}$)= rkG = ℓ by remark 7.7 otherwise $\underline{\bar{X}} = \overline{G/\tilde{H}}$ with G simple.
We know by 7.6 that rk Pic($\underline{\bar{X}}$)> ℓ if and only if there exists a simple root α such that:

$$\alpha^{\sigma} = -\alpha' - \beta \qquad \text{with} \qquad \alpha' = \alpha \quad \text{and either } \beta \neq 0 \text{ or } (\alpha^{\sigma},\alpha') \neq 0.$$

Now we can inspect the tables of Satake diagrams in the classification of symmetric spaces (cf. [10], p. 532-534) and we see using the notations of such tables that the only cases to be considered are the ones denoted by AIII (first diagram) A IV, D III (second diagram), EIII. One remarks by inspecting the table V (p. 518) that these cases belong to table III (p. 515).

In all cases one can verify that there is a unique pair of simple roots α,α' with the above properties and hence rk Pic($\underline{\bar{X}}$)= ℓ+1.

Case AIII and AIV can be explicitely described as follows.

We consider in $\underline{s\ell}_n$ the automorphism σ defined as conjugation by the block matrix

$$\begin{pmatrix} I_k & 0 \\ e & -I_{n-k} \end{pmatrix} \quad \text{with } k \neq n-k.$$

Case DIII can be described as

so(4n+2) relative to the symmetric form

$$\begin{pmatrix} 0 & I_{2n+1} \\ I_{2n+1} & 0 \end{pmatrix}$$

and conjugation relative to

$$\begin{pmatrix} I_{2n+1} & 0 \\ 0 & -I_{2n+1} \end{pmatrix}$$

For E III consider the Dynkin diagram of E_6 indexed as

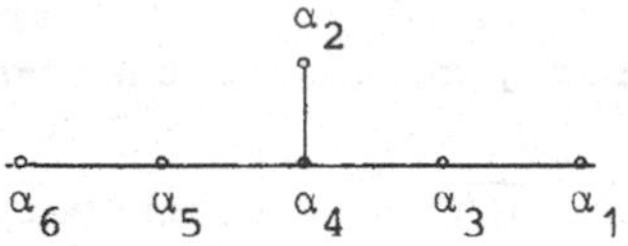

relative to a Cartan subalgebra $\underline{t}$.

Denote by x_α the generator of the corresponding root subspace and define σ as the identity on $\underline{t}$,

$$\sigma(x_{\alpha_i}) = x_{\alpha_i}, \quad i \neq 1, \quad \sigma(x_{\alpha_1}) = -x_{\alpha_1}.$$

One can now verify in each case that the fixed group H is the intersection of a suitable maximal parabolic subgroup Q of type α with its opposite Q' which in all cases is of type α'.

Let us denote by ω and ω' the dual fundamental weights to α, α' and V_ω, $V^*_{\omega'}$ the corresponding irreducible representations. We remark that $V_\omega \simeq V_{\omega'}$ and by 1.3 that $\omega^\sigma = -\omega'$, so that $\omega + \omega'$ is a special weight. If $v \in V_\omega$ (resp. $v' \in V_{\omega'}$) generate the line fixed by Q (resp. by Q') we have that v, v' are seminvariants under H and $v \otimes v'$ is an H invariant, thus if we project $v \otimes v'$ on $V_{\omega+\omega'}$ we obtain a non zero H invariant. By the analysis of section 4 we have a regular morphism π of $\bar{\underline{X}}$ onto the orbit closure Y of the class of $v \otimes v'$ in $\mathbb{P}(V_{\omega+\omega'})$.

We show now that Y is isomorphic to $G/Q \times G/Q'$. This follows from Lemma 8.4 in the following way. In $\mathbb{P}(V_\omega \otimes V_{\omega'})$ the $G \times G$ orbit of $v \otimes v'$ is clearly $G/Q \times G/Q'$ and this orbit projects isomorphically to its image in $\mathbb{P}(V_{\omega+\omega'})$ under $\bar{p}$. On the other hand an easy computation of dimensions shows that the G orbit of $v \otimes v'$ is open in $G/Q \times G/Q'$ hence its closure is $G/Q \times G/Q'$. Since $\bar{p}$ is G-equivariant everything is proved. Comparing the map $\bar{\underline{X}} \to Y \simeq G/Q \times G/Q'$ with the two projections and the respective Plücker embeddings we have two regular projective morphisms associated to the non special weights ω, ω'. We go back now to γ and claim that a suitable positive multiple of γ is of the form $\zeta + a\omega$ or $\zeta + a\omega'$ with $a > 0$ and ζ a dominant special weight.

This can be shown remarking that the subgroup Γ' of Γ generated by the special weights and ω has the same rank as Γ thus a positive multiple of γ lie in Γ'. Now if a dominant weight is in Γ', using the notations of 1.3 it is of the form

$$m\gamma = \sum n_i\omega_i + a\omega \quad \text{with} \quad n_i = n_{\tilde{\sigma}(i)},$$

and ω (resp. ω') is one of the ω_i's, for istance $\omega = \omega_1$ (resp. $\omega' = \omega_2$).

Also $m\gamma$ being dominant $n_1 + a \geq 0$ and $n_i \geq 0$ for $i \neq 1$. If $a \geq 0$ we are done otherwise

$$m\gamma = (n_1 + a)(\omega + \omega') + \sum_{i>2} n_i \omega_i - a\omega'.$$

From this it is clear that for any dominant $\gamma \in \Gamma$ the complete linear system associated to a suitable positive multiple of γ is without base points, since μ is very ample this implies that $\mu + \gamma$ is ample.

9. COMPUTATION OF THE CHARACTERISTIC NUMBERS

9.1. In section 7 we have computed Pic $(\underline{\bar{X}}) \simeq H^2(\underline{\bar{X}},\mathbb{Z})$. We want now to give an explicit algorithm to compute the characteristic numbers. This means that, given n elements $x_1,\ldots,x_n \in H^2(\underline{\bar{X}},\mathbb{Z})$, $n = \dim \underline{\bar{X}}$, we wish to evaluate the product $x_1\ldots x_n \in H^{2n}(\underline{\bar{X}},\mathbb{Z})$ against the class of a point.

Given n reduced hypersurfaces $D_1,\ldots,D_n$ in $\bar{G}/\bar{H}$ such that their closures in $\underline{\bar{X}}$, $\bar{D}_i$ do not contain the unique closed orbit, if $x_i = \mathcal{O}(\bar{D}_i) \in$ Pic $(\underline{\bar{X}}) \simeq H^2(\underline{\bar{X}},\mathbb{Z})$ the corresponding characteristic number counts exactly the number of points common to generic translates g_iD_i, $g_i \in G$, of the D_i's (this is an easy consequence of [12] since $\underline{\bar{X}}$ has a finite number of orbits).

We may work in $H^2(\underline{\bar{X}},\mathbb{Q})$ and use suitable bases for this space. We may also assume that $\underline{\bar{X}}$ is simple (cf. 5.3).

It follows from the analysis performed in section 8 that Pic $(\underline{\bar{X}}) \otimes \mathbb{Q}$ can be identified with the vector space generated by the special weights if $\underline{\bar{X}}$ is not exceptional, otherwise one has to add to the special weights a fundamental weight ω.

Let us denote with Σ the vector space spanned by the special weights and, in the exceptional case $\Gamma_{\mathbb{Q}} = \Sigma + \mathbb{Q}\omega$.

We also know that the divisors S_i correspond to twice the restricted simple roots and form a basis of Σ. Denote by $[S_i]$ these elements in Σ. We have another basis of Σ given by the elements λ_j (cf. 4.1). We notice that $(\lambda_j,[S_i]) = 0$ if $i \neq j$ (for the Killing form).

LEMMA. If $i_1,\ldots,i_k$, $j_1,\ldots,j_{\ell-k}$ is a shuffle of the indices $1,2,\ldots,\ell$, the elements $\lambda_{i_1},\ldots\ldots,\lambda_{i_k},[S_{j_1}],\ldots\ldots,[S_{j_{\ell-k}}]$ form a basis of Σ.

PROOF. Clear by the orthogonality relations.

9.2. Given an oriented compact manifold X and an oriented submanifold Y denote by [Y] the Poincaré dual of the fundamental class of Y. We shall use the following basic facts:

1) If Y_1, Y_2 are oriented submanifolds of X with transversal intersection we have:

$$[Y_1 \cap Y_2] = [Y_1] \cup [Y_2]$$

2) If $Y \subset X$ is a d-dimensional oriented submanifold and $c \in H^d(X)$ we have that the evaluation of $c \cup [Y]$ on the class of a point in X equals the evaluation of $c|_Y$ on the class of a point in Y.

The main proposition is the next one.

PROPOSITION. Let $S_{\{i_1 \ldots i_k\}} = S_{i_1} \cap \ldots \cap S_{i_k}$. If $S_{\{i_1 \ldots i_k\}}$ is not the closed orbit in $\bar{X}$ then:

1) Every monomial $\lambda_{i_1}^{h_1} \lambda_{i_2}^{h_2} \cdots\cdot \lambda_{i_k}^{h_k}$ with $\Sigma h_i = \dim S_{\{i_1 \ldots i_k\}}$ vanishes on $S_{\{i_1 \ldots i_k\}}$.

2) In the exceptional case every monomial $\omega^{h_0} \lambda_{i_1}^{h_1}, \lambda_{i_2}^{h_2} \cdots\cdot \lambda_{i_k}^{h_k}$ with $\Sigma h_i = \dim S_{\{i_1 \ldots i_k\}}$ vanishes on $S_{\{i_1 \ldots i_k\}}$.

PROOF. 1) Recall that we have a projection $\pi\colon S_{\{i_1 \ldots i_k\}} \to G/P_{\{i_1 \ldots i_k\}}$ and the classes $\lambda_{i_1}, \ldots, \lambda_{i_k}$ come via π^* from the cohomology of $G/P_{\{i_1 \ldots i_k\}}$. Since $S_{\{i_1 \ldots i_k\}}$ is not the closed orbit we have $\dim S_{\{i_1 \ldots i_k\}} > \dim G/P_{\{i_1 \ldots i_k\}}$ and everything follows.

2) We have seen in 8.4 that L_ω induces a morphism $p\colon \bar{X} \to G/Q$ for a suitable maximal parabolic Q and ω is the pullback of the ample generator of Pic (G/Q) by p*. We wish to consider the induced map $\pi \times p\colon S_{\{i_1 \ldots i_k\}} \to G/P_{\{i_1 \ldots i_k\}} \times G/Q$ and denote by $\tilde{S}_{\{i_1 \ldots i_k\}}$ its image. We know that $\omega + \omega^\sigma$ is one of the fundamental special weights λ_i. If the index i is one of the indeces of the set $\{i_1, \ldots, i_k\}$ then the parabolic Q contains $P_{\{i_1 \ldots i_k\}}$ and the projection $p\colon S_{\{i_1 \ldots i_k\}} \to G/Q$ factors through $G/P_{\{i_1 \ldots i_k\}}$. This case therefore follows as in 1).
Otherwise $G/P_{\{i_1 \ldots i_k\}} \times G/Q$ contains a unique closed orbit under G isomorphic to $G/P_{\{i_1 \ldots i_k\}} \cap Q$. We claim that $\tilde{S}_{\{i_1 \ldots i_k\}}$ equals this orbit. In fact first of all the fiber of the projection $G/P_{\{i_1 \ldots i_k\}} \cap Q \to G/P_{i_1 \ldots i_k}$ equals the variety $L_{\{i_1 \ldots i_k\}}/L_{\{i_1 \ldots i_k\}} \cap Q$ which is a complete homogeneous space over the semisimple part of $L_{\{i_1 \ldots i_k\}}$.

If we restrict to a fiber $X_{\{i_1 \ldots i_k\}}$ of π the line bundle L_ω we obtain a line bundle of the same type (relative to the minimal compactification $X_{\{i_1 \ldots i_k\}}$ of $\bar{L}_{\{i_1 \ldots i_k\}}/\bar{H}_{\{i_1 \ldots i_k\}}$ (cf. 5.2)).
Since we know that $H^0(X_{\{i_1 \ldots i_k\}}, L_\omega | X_{\{i_1 \ldots i_k\}})$ is an irreducible $L_{\{i_1 \ldots i_k\}}$ module we get that the restriction homomorphism

$$H^0(X, L_\omega) \to H^0(X_{\{i_1 \ldots i_k\}}, L_\omega | X_{\{i_1 \ldots i_k\}})$$

is onto. Hence the induced morphism on $X_{\{i_1 \ldots i_k\}}$ coincides with the restriction to $X_{\{i_1 \ldots i_k\}}$ of p and maps it onto $\bar{L}_{\{i_1 \ldots i_k\}}/L_{\{i_1 \ldots i_k\}} \cap Q$. This proves the claim. Since $S_{i_1 \ldots i_k}$ is not the closed orbit $\dim \tilde{S}_{\{i_1 \ldots i_k\}} < \dim S_{\{i_1 \ldots i_k\}}$ and everything follows as in 1.

9.3. We are now ready to illustrate the algorithm. We treat the exceptional case, the non exceptional is the same without the appearence of ω.

Consider monomials of degree n of type $M = [S_{i_1}] \ldots [S_{i_k}] \omega^{h_0} \lambda_{j_1} \ldots \lambda_{j_s}$ with $i_1, \ldots, i_k$ distinct (in particular the ones with $k = 0$ are the monomials we wish to evaluate). We call k the index of M. We count the number of indices j_h appearing in M and different from $i_1, i_2, \ldots, i_k$ and call this the content of M.

If $j_1 \neq i_1, i_2, \ldots, i_k$ we have an explicit formula expressing λ_{j_1} in terms of $\lambda_{i_1}, \lambda_{i_2}, \ldots, \lambda_{i_k}$ and the $[S_j]$'s relative to the remaining indeces (Lemma 9.1).

Substituting we obtain M expressed as a linear combination of monomials of higher index and of lower content.

Iterating we obtain M as a combination of monomials of index ℓ or of content 0.

By Proposition 9.2 all monomials of contenent 0 vanish, the computation of the remaining ones can be performed:

LEMMA. The evaluation of $[S_1][S_2] \ldots [S_\ell] \omega^{h_0} \lambda_{i_1} \ldots \lambda_{i_k}$ on the class of a point in X equals the evaluation of $\omega^{h_0} \lambda_{i_1} \ldots \lambda_{i_k}$ restricted to the closed orbit on the class of a point in it.

PROOF. Clear since the closed orbit is the transversal intersection of the hypersurfaces S_i.

We summarize

THEOREM. By an explicit algorithm the computation of the characteristic numbers is reduced to the one relative to the closed orbit (for which it is known since the cohomology ring of a complete homogeneous space is known [3]).

10. AN EXAMPLE

10.1. In his fundamental work [14] H. Schubert has computed the number of space quadrics tangent to 9 quadrics in general position to be 666.841.088. We want here to perform again this computation.

The variety of non degenerate quadrics in $\mathbb{P}^n$ is symmetric, it is $X_0 = SL(n+1)/\tilde{SO}(n+1)$ (the involution being $\sigma(A) = {}^tA^{-1}$).

The variety X is classically called the variety of complete quadrics ([1],[15],[17],[19],[21],[22]).

One can easily verify (by the invariant theory of the orthogonal group) that the irreducible representations of SL(n+1) containing an invariant for SO(n+1) are exactly the ones of highest weight $\sum_{i=1}^{n} n_i 2\omega_i$ (ω_i the fundamental weights). From this it follows that we can identify Pic ($\bar{X}$) with 2Λ where Λ is the lattice of weights for SL(n+1) and that the closed orbit in $\bar{X}$ is the full flag variety F. The usual maximal Torus of diagonal matrices is anisotropic and so the restricted simple roots coincide with the usual simple roots. Hence:

$$[S_1] = 2(2\omega_1) - 2\omega_2$$
$$[S_i] = 2(2\omega_i) - 2\omega_{i-1} - 2\omega_{i+1} \qquad 1 < i < n$$
$$[S_n] = 2(2\omega_n) - 2\omega_{n-1}.$$

Let us fix for each $i = 0,\dots,n-1$ a linear subspace π_i of dimension i in $\mathbb{P}^n$. Denote by D_i the hypersurfaces in X_0 of quadric tangent to π_i. We also fix a non degenerate quadric Q and denote by D the hypersurface in X_0 of quadrics tangent to Q. We denote as usual by $\bar{D}_i$, $\bar{D}$ their closures in $\bar{X}$.

PROPOSITION.

1) $[\bar{D}_i] = \mathcal{O}(\bar{D}_i) = L_{2\omega_i}$.

2) $[\bar{D}] = 2\sum_{i=0}^{n-1} [\bar{D}_i]$

3) $\bar{D}_i$ and $\bar{D}$ do not contain the closed orbit.

PROOF. 1) X_0 is the affine variety of symmetric $(n+1)\times(n+1)$ matrices of determinant 1. The map from X_0 to $\mathbb{P}(V^*_{2\omega_i})$ is easily seen to be induced by the map associating to each matrix the matrix of determinants of $i\times i$ minors, which gives a quadric in $\mathbb{P}(V_{\omega_i})$ whose intersection with the Grassmann variety $G_{i-1,n}$ of i-1 dimensional subspaces is exactly the set of tangent subspaces to the original quadric.

Given an i-1 dimensional subspace π_{i-1} in $\mathbb{P}^n$ we consider it as a point in $G_{i-1,n}$, hence, by taking the embedding of $G_{i-1,n}$ in $\mathbb{P}(V_{2\omega_i})$ as a point in $\mathbb{P}(V_{2\omega_i})$. Then it is clear that the intersection of X_0 with the hyperplane in $\mathbb{P}(V^*_{2\omega_i})$ associated to this point is at least set theoretically D_{i-1}. So we have found an $s \in H^0(\bar{X}, L_{2\omega_i})$ whose divisor has support equal to D_{i-1}. But it is clear from our computation of Pic (X) that the divisor of s is reduced so it equals $\bar{D}_{i-1}$ proving 1).

2) Consider the variety $F_{0,n-1}$ of flags $p \in \pi \subset \mathbb{P}^n$ where p is a point and π is an hyperplane. Define a flag (p,π) to be tangent to a quadric $Q \in \bar{X}_0$ if $p \in Q$ and π is the hyperplane tangent to Q in p. Let $Y \subset \bar{X} \times F_{0,n-1}$ be the closure of the correspondence $\tilde{Y} = \{(Q,(p,\pi)) \mid (p,\pi)$ is tangent to Q, $Q \in \bar{X}_0\}$. Clearly $\dim Y = \dim \bar{X} + n - 1 = \frac{(n+1)(n+2)}{2} + n - 2$ and we get two projections

$$\begin{array}{ccc} & Y & \\ {}^{g}\swarrow & & \searrow^{f} \\ \bar{X} & & F_{0,n-1} \end{array}$$

A simple dimension count shows that we have an homomorphism

$$g_* f^* : H^n(F_{0,n-1}, \mathbb{Z}) \to H^2(\bar{X}, \mathbb{Z})$$

Consider our complete flag $\pi_0 \subset \pi_1 \subset \quad \subset \pi_{n-1} \subset \mathbb{P}^n$.
It is well known that a basis of $H^n(F_{0,n-1}, \mathbb{Z})$ is given by the classes dual to the following Schubert subvarieties:

$$Y_i = \{(p,\pi) \mid p \subset \pi_i \subset \pi\}.$$

On the other hand it follows easily from our definition of Y that $g_* f^*([Y_i]) = [\bar{D}_i]$ so that $g_* f^*$ is an isomorphism.

Furthermore if we fix a quadric $Q \in \bar{X}_0$ and we embed it in $F_{0,n-1}$ by associating to each point in Q its tangent flag we get that $g_* f^*([Q]) = [\bar{D}]$ so that in order to prove our claim it is sufficient to show that

$$[Q] = \sum_{i=0}^{n-1} 2[Y_i] \quad \text{in} \quad H^2(F_{0,n-1}, \mathbb{Z})$$

Denote by $Y'_0, \ldots, Y'_{n-1}$ the Schubert cycles dual to $Y_0, \ldots, Y_{n-1}$; i.e. $Y'_i = \{(p,\pi) \mid p \subset \pi_{n-i}, \pi \supset \pi_{n-i-1}\}$. We are reduced to show that the evaluation on the class of a point in $F_{0,n}$ of $[Q] \cdot [Y_i]$ is 2 for each $0 \leq i \leq n-1$. This is clear by elementary considerations on the geometry of quadrics.

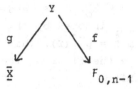

3) We first show that $\bar{D}_i \not\supset F$ for each $0 \leq i \leq n-1$. Assume the contrary and let $s \in H^0(\bar{X}, L_{2\omega_i})$ be a section whose divisor is $\bar{D}_i$. The restriction of s to F is zero. On the other hand it follows from our results of section 8 that the restriction homomorphism

$$j^*: H^0(\bar{X}, L_{2\omega_i}) \to H^0(F, L_{2\omega_i}|F)$$

is an isomorphism.

We now show our result for $\bar{D}$. For this, given a non singular quadric $Q \in X_0$, define a flag $f \in F$ to be tangent to Q if the point of f lies in Q and the hyperplane of f is the hyperplane tangent to Q in this point. Consider the variety $Z \subset \bar{X} \times F$ which is the closure of the correspondence $\tilde{Z} = \{(Q,f) \mid Q \in X_0,\ f \text{ is tangent to } Q\}$. Consider the fibration $p: \bar{X} \times F \to \bar{X} \times F_{0,n-i}$ induced by the natural fibration $q: F \to F_{0,n-i}$. Then we claim $Z = p^{-1}(Y)$. This is clear since $\tilde{Z} = p^{-1}(\tilde{Y})$. This allows us to determine the fiber of the projection $g: Z \to \bar{X}$ over a point f_0 in the closed orbit.

In fact think of f_0 as a flag $f_0 = \{\pi_0 \subset \pi_1 \subset \ldots \subset \pi_{n-1} \subset \mathbb{P}^n\}$ and for each $f \in g^{-1}(f_0)$ put $q(f) = (p,\pi)$. We claim that $g^{-1}(f_0) = \cup Z_i$, where $Z_i = \{f \mid p \subset \pi_i \subset \pi\}$.

To see this notice that the image of f_0 in $\mathbb{P}(V^*_{2\omega_i})$ under the morphism $\bar{X} \to \mathbb{P}(V^*_{2\omega_i})$ represents a degenerate quadric in $\mathbb{P}(V_{\omega_i})$ whose intersection with the Grassmannian of i-1 dimensional subspaces is just the set of such subspaces intersecting π_{n-1}.

Thus if $f \in g^{-1}(f_0)$ its (i-1) dimensional subspace has to meet π_{n-i}. In particular $p \in \pi_{n-1}$.

Assume $p \in \pi_i - \pi_{i-1}$. We claim $\pi \supset \pi_i$. In fact if $i \geq 1$ each (n - i) dimensional subspace τ with $p \in \tau \subset \pi$ has to meet π_{i-1} by the above remarks, and if $i = 0$ there is nothing to prove. So $f \in Z_i$. Having shown this it is easily seen that given f_0 in the closed orbit of $\bar{X}$ such that π_i is not tangent to Q for all $0 \leq i \leq n-1$, $f_0 \notin \bar{D}$ proving 3).

COROLLARY. The evaluation at the class of a point of any monomial of the form

$$(2\omega_1)^{h_1} \cdots\cdots (2\omega_n)^{h_n} \left(2 \sum_{i=1}^{n} 2\omega_i\right)^{h_{n+1}}$$

with $\sum_{i=1}^{n+1} h_i = \frac{(n+1)(n+2)}{2} - 1 = \dim \bar{X}$ gives the number of quadrics which are simultaneously tangent to h_1 points, h_2 lines,....., h_n hyperplanes, h_{n+1} quadrics lying in general position.

REMARK. Our proof of the fact that $\bar{D} \not\supset F$ works also in the case in which $\bar{D}$ is the closure in $\bar{X}$ of the hypersurface of X_o of quadrics tangent to any fixed subvariety in $\mathbb{P}^n$. Thus since $[\bar{D}]$ can be written as a linear combination of the $[\bar{D}_i]$'s the problem of enumerating the number of quadrics simultaneously tangent to $\frac{(n+1)(n+2)}{2} - 1$ subvarieties in general position is reduced to the same problem for linear spaces. This fact has been recently shown in a much greater generality by Fulton, Kleiman, Mac Pherson.

In the case of $\mathbb{P}^3$ working out the computations with the algorithm given in 9.2 one finds the following table which can also be found in Schubert's book (p. 105):

$x_1^9 = x_3^9 = 1$	$x_1^6x_2^2x_3 = x_3^6x_2^2x_1 = 12$
$x_1^8x_2 = x_3^8x_2 = 2$	$x_1^5x_2^3x_3 = x_3^5x_2^3x_1 = 24$
$x_1^7x_2^2 = x_3^7x_2^2 = 4$	$x_1^4x_2^4x_3 = x_3^4x_2^4x_1 = 48$
$x_1^6x_2^3 = x_3^6x_2^3 = 8$	$x_1^6x_2x_3^2 = x_3^6x_2x_1^2 = 18$
$x_1^5x_2^4 = x_3^5x_2^4 = 16$	$x_1^5x_2^2x_3^2 = x_3^5x_2^2x_1^2 = 36$
$x_1^4x_2^5 = x_3^4x_2^5 = 32$	$x_1^4x_2^3x_3^2 = x_3^4x_2^3x_1^2 = 72$
$x_1^3x_2^6 = x_3^3x_2^6 = 56$	$x_1^5x_2x_3^3 = x_3^5x_2x_1^3 = 34$
$x_1^2x_2^7 = x_3^2x_2^7 = 80$	$x_1^4x_2^2x_3^3 = x_3^4x_2^2x_1^3 = 68$
$x_1x_2^8 = x_3x_2^8 = 92$	$x_1^4x_2x_3^4 = 42$
$x_2^9 = 92$	$x_1^2x_2^6x_3 = x_3^2x_2^6x_1 = 104$
$x_1^8x_3 = x_3x_1^8 = 3$	$x_1^3x_2^5x_3 = x_3^3x_2^5x_1 = 80$
$x_1^7x_3^2 = x_3^2x_1^7 = 9$	$x_1^3x_2^4x_3^2 = x_3^3x_2^4x_1^2 = 112$
$x_1^6x_3^3 = x_3^3x_1^6 = 17$	$x_1x_2^7x_3 = 104$
$x_1^4x_3^5 = x_3^5x_1^4 = 21$	$x_1^2x_2^5x_3^2 = 128$
$x_1^7x_2x_3 = x_3^7x_2x_1 = 6$	$x_1^3x_2^3x_3^3 = 104$

and so

$$(2(x_1 + x_2 + x_3))^9 = 666.841.088$$

REFERENCES

[1] A.R. ALGUNEID: Complete quadric primals in four dimensional space. Proc. Math. Phys. Soc. Egypt, 4, (1952), 93-104.

[2] BIALYNICKI-BIRULA: Some theorems on actions of algebraic groups. Ann. of Math., 98, 1973, 480-497.

[3] A. BOREL: Sur la cohomologie des espaces fibrés principaux et des espaces homogênes de groupes de Lie compacts. Ann. of Math., 57, 1953, 116-207.

[4] R. BOTT: Homogeneous vector bundles. Ann. of Math. 66, 1957, 203-248.

[5] M. DEMAZURE: Limites de groupes orthogonaux ou symplectiques. Preprint 1980.

[6] G. GHERARDELLI: Sul modello minimo delle varietà degli elementi differenziali del 2° ordine del piano proiettivo. Rend. Acad. Lincei, (7) 2, 1941, 821-828.

[7] G.H. HALPHEN: Sur la recherche des points d'une courbe algégrique plane. In "Journal de Mathématique", 2, 1876, 257.

[8] HARISH-CHANDRA: Spherical functions on a semisimple Lie group I. Amer. J. of Math., 80, 1958, 241-310.

[9] S. HELGASON: A duality for symmetric spaces with applications to group representations. Advances in Math. 5, 1-154, (1970).

[10] S. HELGASON: Differential geometry, Lie groups, and symmetric spaces. Acad. Press 1978.

[11] S. KLEIMAN: Problem 15. Rigorous foundation of Schubert enumerative calculus. Proceedings of Symp. P. Math. 28, A. M. S., Providence (1976).

[12] S. KLEIMAN: The transversality of a general translate. Comp. Math., 28, 1974, 287-297.

[13] D. LUNA, T. VUST: Plongements d'espaces homogenès.
Preprint.

[14] H. SCHUBERT: Kalkül der abzählenden geometrie.
Liepzig 1879 (reprinted Springer Verlag 1979).

[15] J.G. SEMPLE: On complete quadrics I.
J.London Math. Soc. 23, 1948, 258-267.

[16] J.G. SEMPLE: The variety whose points represent complete collinea tions of S_r on S'_r.
Rend. Mat. 10, 201-280 (1951).

[17] J.G. SEMPLE: On complete quadrics II.
J. London M. S. 27, 280-287 (1952).

[18] F. SEVERI: Sui fondamenti della geometria numerativa e sulla teo ria delle caratteristiche.
Atti del R. Ist. Veneto, 75, 1916, 1122-1162.

[19] F. SEVERI: I fondamenti della geometria numerativa.
Ann. di Mat., (4) 19, 1940, 151-242.

[20] E. STUDY: Uber die geometrie der kegelschnitte, insbesondere dere charakteristiken problem.
Math. Ann., 26, 1886, 51-58.

[21] J.A. TYRELL: Complete quadrics and collineations in S_n.
Mathematika 3, 69-79 (1956).

[22] I. VAISENCHER: Schubert calculus for complete quadrics.
Preprint.

[23] B.L. VAN DER WAERDEN: Z.A.G. XV, Losung des charakteristiken-problem für kegelschnitte.
Math. Ann. 115, 1938, 645-655.

[24] J. VUST: Opération des groupes réductifs dans un type de cônes presque homogènes.
Bull. Soc. Math. France, 102, 1974, 317-333.

[25] H.G. ZEUTHEN: Abzählende methoden der geometrie.
Liepzig 1914.

[26] A. BIALYNICKI-BIRULA: Some properties of the decomposition of algebraic varieties determined by actions of a torus.
Bull. Acad. Polon. Sci. Ser. Sci. Math. Astronom.
Phys. 24 (1976) n. 9, 667-671.

[27] R. STEINBERG: Générateurs, relations et revêtements de groupes algébriques, p. 113-127, Collq. Theorie des Groupes Algébriques, Gauthier Villars (1962).

[28] R. STEINBERG: Endomorphisms of linear algebraic groups, Mem. of the A.M.S. n. 80 (1968).

GEOMETRIC INVARIANT THEORY AND APPLICATIONS TO MODULI PROBLEMS

D. Gieseker
University of California
Los Angeles, California 90024 USA

These notes are a brief introduction to geometric invariant theory (GIT) and also contain two applications of that theory to the construction of moduli spaces in algebraic geometry. The first two sections sketch the basics of GIT over the complex numbers. In §3 we connect GIT and the theory of stable bundles of rank two on a non-singular curve. We then consider in §4 the relation between smooth curves and GIT. Our main result here is that there are enough projective invariants of space curves to separate any two projectively distinct smooth curves of genus g and degree d provided $d \geq 2g$ and that the curves are non-degenerate. This result can be used to construct a moduli space $\mathfrak{m}_g$ for smooth curves of genus g. In sections §5 and §6, we look at the connection between stable curves in the sense of Mumford and Deligne and stable curves in the senses of GIT. The main result is essentially that the compactification $\overline{\mathfrak{m}_g}$ of $\mathfrak{m}_g$ considered by Mumford and Deligne is a projective variety. (This result was originally obtained by F. Knutsen in characteristic zero using other methods.) Finally in §7 we indicate how GIT can be used to construct compactified generalized Jacobians of stable curves. Here we consider the example of an irreducible curve with one node. The nature of the compactification of the generalized Jacobian of a general stable curve obtained by GIT has yet to be worked out. One can also extend the results of §5,6,7 to vector bundles of rank two [G - M,G_4]. Roughly, one gets a construction of a projective moduli space of stable bundles on an irreducible curve which has one node. This can then be used to study the topology of the moduli space of stable bundles on a smooth curve by degeneration methods.

The original source for the first two sections is [M_1], but [N] also provides a more leisurely treatment. A connection between GIT and the theory of stable bundles on a smooth curve was worked out by Mumford and Seshadri. [N] contains an account of this work. In these notes, we make a slightly different connection which is more suitable for higher dimensional varieties [G_1,Ma]. Mumford gave a proof of the existence of $\mathfrak{m}_g$ using GIT in [M_1] using the Chow variety of a space curves. Here we use Grothendieck's Hilbert scheme which is arguably easier. [G_2] contains an extension of these ideas to the n canonical images of surfaces of general type. The connection between GIT and stable curves was worked out jointly by Mumford and myself using the Chow variety and Hilbert scheme [M_2,G_3]. Finally an exhaustive discussion of the developments in GIT since the first edition of Mumford's book and the present can be found in the second edition of Mumford's book.

§1. Let k be an algebraically closed field and let W be a vector space of dimension n. Let G be the algebraic group $SL(W,k)$. Suppose that V is a vector space of dimension ℓ and that G acts on $V^* = Hom(V,k)$. If $x \in V^*$ and $\sigma \in G$, we will denote the action of σ on x by x^σ. G acts on $\mathbb{P}(V)$, the hyperplanes in V. We would like to be able to form a reasonable quotient of $\mathbb{P}(V)$ by G as a projective variety. Unfortunately, it is usually impossible to form such a quotient. At least, one would hope for a map $m : \mathbb{P}(V) \to \mathbb{P}^K$ which is G invariant and separates orbits. As the following example shows, no such map can exist in general.

Example:

Let $G = SL(3)$ and let $V^* = S^3(k^3)$ with the natural action of G. Choose a $\lambda \in k$, $\lambda \neq 0,1$. For each $t \in k$, define a element of V^* by

$$P_t(X_0,X_1,X_2) = X_0^2X_2 - X_1(X_1 - tX_2)(X_1 - t\lambda X_2)$$

where X_0,X_1,X_2 is a basis for k^3. Let $\overline{P}_t \in \mathbb{P}(V)$ denote the point corresponding to P_t. Note that all the $\overline{P}_t$ are in the same G orbit as $\overline{P}_1$ if $t \neq 0$, but that $\overline{P}_0$ is not in the same G orbit as $\overline{P}_1$. Indeed $\overline{P}_1$ defines a smooth elliptic curve in $\mathbb{P}^2$, but $\overline{P}_0$ defines a cubic curve with a cusp. There is a morphism $F : \mathbb{A}^1 \to \mathbb{P}(V)$ so that $F(t) = \overline{P_t}$. Suppose there is a G invariant map m of $\mathbb{P}(V)$ to $\mathbb{P}^K$ for some N which separates G orbits. Then $m \circ F$ is constant on $\mathbb{A}^1 - \{0\}$ so $m(F(1)) = m(F(o))$. So no such map can exist.

The idea behind GIT is to find a large G invariant open set U of $\mathbb{P}(V)$ so that one can form a reasonable quotient of U by G. For instance in the above example, one must either exclude $\overline{P}_0$ or $\overline{P}_1$ from U. Since an arbitrary smooth elliptic curve is projectively equivalent to $\overline{P}_1$ for some appropriate λ, one sees that one must exclude $\overline{P}_0$ from any U.

Let $x \in V^* - \{0\}$ and let $\overline{x}$ be the corresponding point in $\mathbb{P}(V)$. We define $\rho_x : G \to V^*$ by $\rho_x(\sigma) = x^\sigma$. It is a fact that ρ_x is proper if and only if $x^G = \rho_x(G)$ is closed and the stabilizer of x is finite [N, Lemma 3.17]. Here are the the fundamental definitions of GIT:

Definition 1.1:

1) $\overline{x}$ is stable if ρ_x is proper .
2) $\overline{x}$ is semi-stable if $0 \notin \overline{x^G}$.
3) $\overline{x}$ is weakly stable if x^G is closed.

Here $\overline{x^G}$ is the closure of orbit of x in V^*. Note that we are interested in the orbit of x in V^*, not in the orbit of $\overline{x}$ in $\mathbb{P}(V)$. Let $U_{S.}$ (resp. $U_{S.S.}$) be the set of stable (resp. semi-stable) points. Note that there is much confusion in the literature over these definitions. In particular, weakly stable is not standard terminology.

We wish to define a map $m : U_{S.S.} \to \mathbb{P}^K$ enjoying various pleasant properties.

The method of GIT is to choose a large N and look at a basis $P_0,\ldots,P_K$ of the homogeneous polynomials of degree N which are invariant under G. We then define

$$m(\bar{x}) = (P_0(x),\ldots,P_K(x)) \in \mathbb{P}^K .$$

Of course m is not defined if all the P_i vanish at x. The main result of GIT is the following:

Theorem 1.2:

For N sufficiently large, the map m is defined exactly on $U_{S.S.}$. Further if $\bar{x}$ and $\bar{x}'$ are weakly stable and $m(\bar{x}) = m(\bar{x}')$, then $\bar{x}$ and $\bar{x}'$ are in the same G orbit. Also if $X \subseteq U_{S.S.}$ is closed and G invariant, then $m(X)$ is closed in $\mathbb{P}^K$.

The following corollary is a frequently used consequence of the last statement of Theorem 1.2.

Corollary 1.2.1:

Let S be a smooth curve, $P \in S$, and suppose f is a map of $S - P$ to $U_{S.S.}$. Then there is a curve S', a map $\pi : S' \to S$ and a point $Q \in S'$ so that $\pi(Q) = P$, and a map f' of $S' \to U_{S.S.}$ and a map $h : S' - Q \to G$ so that for $x \in S' - Q$,

$$f(\pi(x)) = (f'(x))^{h(x)} .$$

We may further choose f' so that $f'(Q)$ is stable or has stabilizer of positive dimension.

This corollary is often referred to as the semi-stable replacement property of GIT.

One can say more about the quotient of $U_{S.S.}$. The reader is referred to [N]. The proof of Theorem 1.2 is due to Mumford in characteristic zero. In characteristic p, Nagata first reduced the problem to a conjecture of Mumford. This conjecture was then established by Haboush and independently by Formenck and Procesi.

In these notes we will content ourselves with sketching the proof of Theorem 1.2 when $k = \mathbb{C}$. To start the proof, we put a positive definite hermetian inner product on W, and let $U \subseteq G = SL(W)$ be the special unitary group. The first observation is that U is Zariski dense in G. Indeed, let G' be the Zariski closure of U. The tangent space of G' at the identity contains the complexified tangent space of U. But the tangent space of U consists of traceless matrices with $A = -\overline{A^t}$, where — denotes conjugation. The $\mathbb{C}$ span of the tangent space of U is the tangent space to $SL(n,\mathbb{C})$. Thus $\dim G' = \dim G$, since all algebraic groups are non-singular over $\mathbb{C}$. Thus $G' = G$.

Suppose G acts linearly on a vector space V_1. We claim there is a U invariant positive definite hermetian form on V_1. Let V_2 be the vector space of forms on V_1, linear in the first variable and conjugate linear in the second. Let $h \in V_2$ be a positive definite Hermetian form. Let $B \subseteq V_2$ be the convex hull of the orbit of h under U. B is compact and U invariant, and every element of B is positive definite Hermetian. Now the centroid of B is invariant under U and gives the desired inner product.

Next we note that a subspace $V' \subseteq V$ is G invariant if and only if it is U invariant. Indeed, the stabilizer of V' is a Zariski closed subset of G.

Next we come to the Reynolds operator. Note that V is a direct sum of irreducible representations using the U invariant hermitian inner product. Let V^{inv} be the invariant vectors in V. Then we can construct a G invariant projection from V to V^{inv}. Now let $R_m : S^m(V) \to S^m(V)^{\text{inv}}$ be this projection. Let $g \in S^n(V)$ be invariant. We have the following Reynold's identity: If $f \in S^m(V)$, (1.2.2) $R_{n+m}(fg) = gR_m(f)$. Indeed, define

$$F(f) = R_{n+m}(fg) - gR_m(f).$$

Then F is a G invariant map from $S^m(V)$ to $(S^{m+n}(V))^{\text{inv}}$ which kills $S^m(V)^{\text{inv}}$. On the other hand, if $V_1 \subseteq S^m V$ is a non-trival irreducible subspace, the map of V_1 to $S^{n+m}(V)^{\text{inv}}$ is trivial. So F is trivial. So (1.2.2) is established.

Note every invariant homogeneous polynomial P of positive degree must vanish on all points which are not semi-stable. Indeed, P is constant on any orbit and hence on the closure of any orbit. So such a polynomial must vanish if $0 \in \overline{xG}$.

Let's check that if $x \in V^*$ is semi-stable, then there is a k and a $P \in S^k(V)^G$ so that $P(x) \neq 0$. First, note that since $\{0\}$ and $\overline{xG}$ are closed disjoint sets, there is a polynomial P which vanishes at 0 and which is identically one on $\overline{xG}$. (The ideals I_1 and I_2 of $\{0\}$ and $\overline{xG}$ in $\mathbb{C}[V]$ generate $\mathbb{C}[V]$ by the Nullstellensatz.) Next suppose I_1 and I_2 are invariant ideals in $\mathbb{C}[V]$ and let $\overline{f}_i \in \mathbb{C}[V]/I_i$ be invariant. The if $I_1 + I_2 = \mathbb{C}[V]$, we can find an invariant $f \in \mathbb{C}[V]$ so that f maps to $\overline{f}_i \in \mathbb{C}[V]/I_i$. Indeed, let $\mathbb{C}[V]_n$ be the polynomials of degree $\leq n$. We have a map

$$\varphi_n : \mathbb{C}[V]_n \to \mathbb{C}[V]_n/(I_1 \cap \mathbb{C}[V]_n) \oplus \mathbb{C}[V]_n/(I_2 \cap \mathbb{C}[V]_n).$$

We choose n so that $\overline{f}_i \in \mathbb{C}[V]_n/I_i \cap \mathbb{C}[V]_n$ and so that $(\overline{f}_1, \overline{f}_2)$ is in the image of φ_n. But then there is an invariant $f \in \mathbb{C}[V]_n$ so that $\varphi_n(f) = (\overline{f}_1, \overline{f}_2)$. Hence there is an invariant polynomial f which vanishes at 0 but which is identically one on $\overline{xG}$. Now write $f = \sum_{i=0}^{n} f_i$ as a sum of its homogeneous parts. Thus $f_0 = 0$, but some $f_i(x) \neq 0$. Similarly, one shows that any two weakly stable orbits can be separated by an invariant polynomial.

Next we claim that $C[V]^G$ is finitely generated. Let $R = \oplus R_m$ be the Reynolds operator. Let I be the ideal in $C[V]$ generated by homogeneous invariant polynomials of positive degree. Let $P_1,\ldots,P_m \in I$ be generators of I as a $C[V]$ module, where the P_i are homogeneous and invariant. We claim that $\{1,P_1,\ldots,P_m\}$ generate $C[V]^G$ as a C algebra. We assume inductively that $1,\ldots,P_m$ generate $C[V]^G$ in degree $< n$. Given P homogeneous and invariant of degree n, we can write

$$P = \sum Q_i M_i$$

where the Q_i are monomials of degree $< n$ and the M_i are monomials in $P_1,\ldots,P_m$. Now

$$P = R(P) = \sum R(Q_i) M_i$$

From the induction hypothesis, $R(Q_i)$ are polynomials in $1,\ldots,P_m$. So we can choose an N and $P_0,\ldots,P_K \in S^N(V)^G$ so that $1,P_0,\ldots,P_K$ generate the ring $\oplus S^{Nk}(V)^G = R'$. Let $C[Y_0,\ldots,Y_K]$ be a polynomial ring and consider the surjection

$$\psi : C[Y_0,\ldots,Y_K] \to R'$$

Let $X \subseteq U_{S.S.}$ be closed and G invariant. Let I_X be the ideal of functions in $C[V]^G$ vanishing on X, let $m : U_{S.S.} \to P^K$ be the map defined by $P_0,\ldots,P_K$, and let $I_1 = \psi^{-1}(I_X \cap R')$. Then $\overline{m(X)}$ is defined by the ideal I_1. If $\overline{m(X)} \neq m(X)$, there is a homogeneous ideal $I_2 \supseteq I_1$ so that I_2 maps to a nontrivial ideal I_3 in R', but so that for each point $Q \in U_{S.S.}$, some element of I_3 does not vanish at Q. Let I_4 be the ideal of functions in $C[V]$ vanishing on the non-semistable points. Then $I_4 \subseteq \sqrt{I_3 . C[V]}$. Using the Reynolds operator, we see that

$$M \subseteq \sqrt{I_3}$$

where M is the ideal generated by homogeneous functions of positive degree in R'. This contradicts the nontriviality of I_3.

§2. It is difficult to check directly that a point $x \in V^*$ is stable. However, there is a convenient test called the numerical critereon. Recall that $\mathbb{G}_m$ is just the multiplicative group k^* as an algebraic group. By definition, a one parameter subgroup of G (1-P-S) is a non-trivial map $\lambda : \mathbb{G}_m \to G = SL(W)$. It is well known that there is then a basis $e_1, \ldots, e_n$ of W and integers r_i so that

$$e_i^{\lambda(\alpha)} = \alpha^{r_i} e_i .$$

We will say that e_i has λ weight r_i. Of course, $\sum r_i = 0$, since $G = SL(W)$. (Over $\mathbb{C}$, it is not difficult to see that W decomposes into one dimensional subspaces using unitary techniques.) If $\mu : \mathbb{G}_m \to \mathbb{A}^n = k^n$ is any map, we will use the symbolism $\lim_{\alpha \to 0} \mu(\alpha)$. If μ extends to a map of $\mathbb{A}^1 \supseteq \mathbb{G}_m$ to W, we define

$$\lim_{\alpha \to 0} \mu(\alpha) = \mu(0)$$

Otherwise, we say $\lim_{\alpha \to 0} \mu(\alpha) = \infty$. This means that at least one component of μ goes to ∞ as $\alpha \to 0$. We say $x \in V^*$ is λ-stable (resp. λ semi-stable) if $\lim_{\alpha \to 0} x^{\lambda(\alpha)} = \infty$ (resp. $\lim_{\alpha \to 0} x^{\lambda(\alpha)} \neq 0$).

<u>The numerical criteron</u>: x is stable (resp semi-stable) if and only if x is λ stable (resp. λ semi-stable) for all 1-P-S of G.

<u>Proof</u>:

Let $D = \operatorname{Spec} k[[t]]$ and let $D^* = \operatorname{Spec} k((t))$. Suppose $x \in V^*$ is not stable. Consider the orbit map $\rho_x : G \to V^*$. (One can regard D and D^* as the unit disk and punctured unit disk over $\mathbb{C}$.) By the valuative critereon for properness, we can find a map $\mu : D^* \to G$ which does not extend to a map of D to G, but so that $\rho_x \circ \mu$ does extend to a map of D to V^*. Let $K = k((t))$ and $\mathcal{O} = k[[t]]$. Then μ is equivalent to an element of $SL(n,K)$. Using the theory of elementary divisors, we can find σ_1 and σ_2 in $SL(n,\mathcal{O})$ so that

$$\sigma_1 \mu \sigma_2 = \begin{pmatrix} t^{r_1} & & 0 \\ & \ddots & \\ 0 & & t^{r_n} \end{pmatrix} = \lambda(t)$$

Using $\sigma_1(0)$ to change basis in k^n, we may further assume that $\sigma_1(0)$ is the identity. We claim that $\lim_{\alpha \to 0} x^{\lambda(t)}$ exists. Let $v_1, \ldots, v_\ell$ be a basis of V^* so that

$$v_i^{\lambda(t)} = t^{s_i} v_i .$$

We may assume that $s_1 \leq s_2 \leq \ldots$. Now write $x = \sum x_j v_j$, and assume $x_i \neq 0$ but

$x_j = 0$ for $j < i$. We can write

$$v_i^{\sigma_1^{-1}(t)} = \sum a_{ij}(t) v_j .$$

Thus $a_{ij}(0) = \delta_{ij}$. So

$$\begin{aligned} x^{\mu(t)\sigma_2(t)} &= \sum_i x_i v_i^{\mu(t)\sigma_2(t)} \\ &= \sum_i x_i v_i^{\sigma_1^{-1}(t)\lambda(t)} \\ &= \sum_j \left(\sum_i a_{ij}(t) x_i \right) v_j^{\lambda(t)} \\ &= \sum_j \left(\sum_i a_{ij}(t) x_i \right) t^{s_j} v_j \end{aligned}$$

Now $\lim_{t \to 0} x^{\mu(t)\sigma_2(t)}$ is defined, so if $s_j < 0$, we must have $\sum_i a_{ij}(0) x_i = x_j = 0$. Thus $\lim_{t \to 0} x^{\lambda(t)}$ exists.

A similar argument shows that if x is not semi-stable, then there is a 1 - P - S λ with $\lim_{t \to 0} x^{\lambda(t)} = 0$.

§3. We wish to connect geometric invariant theory to the theory of stable bundles on a smooth curve X. For simplicity we will consider bundles of rank two. Fix a line bundle L of degree $d \gg g$. Let S be the set of bundles E of rank two with $\overset{2}{\Lambda} E \cong L$. Recall that a bundle E of rank two is stable if for every quotient line bundle $E \to L \to 0$, we have

$$\deg L > \frac{\deg E}{2} .$$

Notice first that if $E \in S$ is stable, then E is generated by global sections and $H^1(E) = 0$. Indeed, recall that if s is a section of a bundle F on X, then there is subbundle $F' \subseteq F$ of rank one so that $s \in H^0(F') \subseteq H^0(F)$. In particular $\deg F' \geq 0$. Thus if $H^1(E) \neq 0$, Serre duality shows that $H^0(E^{-1} \otimes \Omega^1) \neq 0$, so $E^{-1} \otimes \Omega^1$ has a sub bundle of non-negative degree. So E has a quotient bundle of degree $\leq 2g - 2$. This contradicts our assumptions that $d \gg g$ and that E is stable. Similary $H^1(E(-P)) = 0$ for any point $P \in X$, so E is generated by global sections.

We next associate to each stable $E \in S$ a point in some projective space modulo the action of an algebraic group. We fix a basis $s_1, \ldots, s_k$ of $H^0(E)$, where $k = d + 2(1 - g)$. Such a basis gives an isomorphism φ of $H^0(E)$ with C^k. There is a canonical map

$$\overset{2}{\Lambda} H^0(E) \to H^0(\overset{2}{\Lambda} E) .$$

obtained by sending $s_1 \wedge s_2$ to the corresponding section in $\overset{2}{\Lambda} E$. We have assumed that $\overset{2}{\Lambda} E \cong L$, so there is an canonical isomorphism of $H^0(\overset{2}{\Lambda} E)$ with $H^0(L) = V$ determined up to scalar multiple. On the other hand, φ determines an isomorphism of $\overset{2}{\Lambda} H^0(E)$ with $\overset{2}{\Lambda} C^k$. Putting these together, we obtain a map depending on E and φ:

$$T_{E,\varphi} \in \mathrm{Hom}(\overset{2}{\Lambda} C^k, V) .$$

We claim that if $T_{E,\varphi} = T_{F,\varphi'}$ and E is generated by global sections, then $E \cong F$. Indeed, since E is generated by global sections, there is a map from X to G', the grassmannian of all two dimensional quotients of C^k. Via the Plücker coordinates, G' is embedded in $\mathbb{P}(\overset{2}{\Lambda} C^k)$. In fact, the induced map of X to $\mathbb{P}(\overset{2}{\Lambda} C^k)$ is given by the linear system generated by $s_i \wedge s_j$ as sections of $H^0(\overset{2}{\Lambda} E) = H^0(L)$. So $T_{E,\varphi}$ determines E. Thus isomorphism classes of bundles E generated by global sections with $H^1(E) = 0$ map injectively to the orbits of $\mathbb{P}(\mathrm{Hom}(\overset{2}{\Lambda} C^K, V))$ under $SL(k,C)$.

We claim that if E is a stable bundle, then $T_{E,\varphi}$ is stable in the sense of GIT. Suppose that $T_{E,\varphi}$ is not stable. There is a $1-P-S$ λ so that $T_{E,\varphi}$ is not λ-stable. Let v_i be a basis of $H^0(E) \overset{\varphi}{=} C^k$ so that $v_i^{\lambda(\alpha)} = \alpha^{r_i} v_i$. Thus $v_i \wedge v_j = 0$ in $H^0(\overset{2}{\Lambda} E)$ if $r_i + r_j < 0$. Order the r_i so that $r_1 \leq r_2 \leq \cdots$. Let M be the subline bundle of E so that v_1 is a section of M. Note that $v_1 \wedge v_j = 0$ if $r_1 + r_j < 0$. Thus $v_j \in H^0(M)$ if $r_1 + r_j < 0$. $N = \#\{j \mid r_j < |r_1|\}$. Then $h^0(M) \geq N$. If $h^1(M) \neq 0$, then $h^0(M) \leq g$, while if $h^1(M) = 0$, $h^0(M) = \deg M + 1 - g < \frac{1}{2}(\deg E + 2(1-g)) = k/2$, since we have assumed E is stable. Thus $N < k/2$. For more than $\frac{k}{2}$ of the j, we have $r_j > |r_1|$. This contradicts $\sum r_i = 0$.

One can continue this line of reasoning to produce a quasi-projective moduli space for stable bundles, even over smooth varieties of higher dimension [G_1,Ma].

§4. Our object in this section is to construct a moduli space for smooth curves of genus g. This was first done by GIT by Mumford $[M_1, M_2]$ using Chow points. Here we will work with the Hilbert point of a curve. Full details may be found in $[G_3]$.

Let $X \subseteq \mathbb{P}(W)$ be a closed subscheme. We let $P_X(n) = \chi(\mathcal{O}_X(n))$. Recall that P_X is a polynomial in n, called the Hilbert polynomial of X. Let P be a fixed polynomial and consider the set U of subschemes of $\mathbb{P}(W)$ with Hilbert polynomial P. We will assign to each $X \in U$ a point $H_m(X)$ in some projective space $\mathbb{P}^M$ which will classify X as a subscheme of $\mathbb{P}(W)$. Further $G = SL(W,k)$ will operate on $\mathbb{P}^M$ and two subschemes X and X' will be projectively equivalent if and only if $H_m(X)$ and $H_m(X')$ are in the same orbit. We will then investigate the stability of $H_m(X)$. We will say X is m-stable (or m-Hilbert stable) if $H_m(X)$ is stable. The main result of invariant theory can be phrased as follows: There are enough projective invariants of subschemes of $\mathbb{P}(W)$ to separate any two m-stable subschemes which are not projectively equivalent.

One case which is of particular interest is that of n canonically embedded curves. Let $\dim W = (2n-1)(g-1)$ and let X be a smooth projective curve of genus g. Choosing an isomorphism of W with $H^0(X, \mathcal{O}(nK))$ determines an embedding of X into $\mathbb{P}(W)$ if $n \geq 2$. One of the main results of these notes will be that this embedding is m-stable for $m \gg 0$. Notice that the n canonical images of two curves are projectively equivalent if and only if the curves are isomorphic. Thus the projective invariants of an n canonical curve form moduli for the curve.

We turn to Grothendieck's construction of the mth Hilbert point $H_m(X)$. Let $X \in U$. Note $H^0(\mathbb{P}(W), \mathcal{O}(m)) = S^m(W)$. There is a natural map

$$\psi : S^m(W) \to H^0(X, \mathcal{O}_X(m))$$

The elements of $S^m(W)$ are just homogeneous polynomials of degree m on $\mathbb{P}(W)$, and ψ is just the restriction map. The kernel of ψ consists in the homogeneous polynomials which vanish on X. According to fundamental results of Grothendieck, there is an M depending only on P so that for $m \geq M$, ψ is surjective , the kernel of ψ determines X, and $\dim H^0(X, \mathcal{O}(m)) = P(m)$. Consider

$$\psi' = \overset{P(m)}{\Lambda} \psi : \overset{P(m)}{\Lambda} S^m(W) \to \overset{P(m)}{\Lambda} H^0(X, \mathcal{O}(m)) \cong k.$$

Then ψ' is onto and hence determines a point $H_m(X)$ in $\mathbb{P}(\overset{P(m)}{\Lambda} S^m(W))$. Note that $H_m(X)$ determines $\overset{P(m)}{\Lambda} \psi$ up to a scalar. Further, $\overset{P(m)}{\Lambda} \psi$ are the Plücker coordinates of $\ker \psi$. Note finally that $\ker \psi$ determines X. Hence $H_m(X)$ determines X.

We will now discuss the stability of $H_m(X)$ under G. Let λ be a 1 - P - S of G and let $X_0, \ldots, X_{n-1}$ be a basis of W so that

$$X_i^{\lambda(\alpha)} = \alpha^{r_i} X_i \ .$$

We will say that X_i has λ weight r_i. A monomial $M = X_0^{i_0} \ldots X_{n-1}^{i_{n-1}}$ will be said to have λ weight $\sum i_j r_j$. The λ weight of M is denoted $w_\lambda(M)$.

Proposition 4.1.

$H_m(X)$ is λ-stable if and only if there are monomials $M_1, \ldots, M_{P(m)}$ in $S^m(W)$ so that the following two conditions hold:

i) $\psi(M_1), \ldots, \psi(M_{P(m)})$ are a basis for $H^0(X, \mathcal{O}(m))$.

ii) $\sum w_\lambda(M_i) < 0$.

Proof:

$$\left(\overset{P(m)}{\Lambda} \psi\right)^{\lambda(\alpha)} (M_1 \wedge \ldots \wedge M_{P(m)}) = \alpha^{\sum w_\lambda(M_i)} \psi(M_1) \wedge \ldots \wedge \psi(M_{P(m)}).$$

For $H_m(X)$ to be λ stable, there must exist monomials $M_1, \ldots, M_{P(m)}$ so that $\sum w_\lambda(M_i) < 0$ and $\psi(M_1) \wedge \ldots \wedge \psi(M_{P(m)}) \neq 0$.

We can obtain a critereon for semi-stability by substituting (ii bis) $\sum \omega_\lambda(M_i) \leq 0$ for condition (ii).

Next, we suppose X is a smooth curve of degree d and genus g. We suppose $X \subseteq \mathbb{P}(W)$ embedded by a complete linear system and that the genus of X is at least one.

Our aim is to sketch a proof of the following result:

Theorem 4.2.

For $m \gg 0$, $H_m(X)$ is stable if

i) X is not contained in any hyperplane

ii) $d \geq 2g$.

Theorem 4.2 is essentially due to Mumford [M_1] except that Mumford considers the stability of the Chow point of X rather than the mth Hilbert point of X. More precisely, we should show that there is an M depending only on d and g so that if $m \geq M$, and conditions (i) and (ii) are satisfied, then $H_m(X)$ is stable.

Let λ be a 1 - P - S and let X_i be a basis of $H^0(X, \mathcal{O}_X(1)) = W$ so that $X_i^{\lambda(\alpha)} = \alpha^{r_i} X_i$ with $r_1 \leq r_2 \leq \ldots \leq r_\ell$. Our aim is to examine the λ-stability of $H_m(X)$. To this end, we let F_i be the subsheaf of $\mathcal{O}_X(1)$ generated by $X_1, \ldots, X_i$. Thus F_i is the smallest subsheaf of $\mathcal{O}_X(1)$ so that $X_j \in H^0(X, F_i) \subseteq H^0(X, \mathcal{O}_X(1))$ for $j \leq i$. Now the F_i are line bundles on X. Let $e_i = \deg F_i$.

Thus $e_1 = 0$, since $F_1 = \mathcal{O}_X$ and $e_\ell = d$, since $\mathcal{O}_X(1)$ is generated by $X_1,\ldots,X_\ell$.

Proposition 4.3.

Suppose there are integers i_j with $1 = i_1 < \ldots < i_k = \ell$ so that

$$\sum_f \frac{e_{i_f} + e_{i_{f+1}}}{2}\left(r_{i_{f+1}} - r_{i_f}\right) > r_\ell d .$$

Then $H_m(X)$ is λ stable if $m = (L+1)N$ with $N \gg L \gg 0$.

Proof.

If L_1 and L_2 are two line bundles on X and V_i is a subspace of $H^0(X,L_i)$, we let $V_1.V_2$ denote the subspace of $H^0(X,L_1 \otimes L_2)$ spanned by elements of the form $v_1 \otimes v_2$, $v_i \in V_i$. With this notation, for any L and $0 \leq p \leq L$ consider

$$V'_{i,j} = V_i^{L-p}.V_j^p \subseteq H^0(X,F_i^{L-p} \otimes F_j^p).$$

Notice that the sections in $V'_{i,j}$ generate the sheaf $F_i^{L-p} \otimes F_j^p$. V_ℓ is a very ample linear system. Hence $V'_{i,j}.V_\ell = V_{i,j}$ is a very ample linear system.

$$V_{i,j} \subseteq H^0(X,\ F_i^{(L-p)} \otimes F_j^p \otimes \mathcal{O}_X(1))$$

Fix i,j,L and p momentarily and let

$$M = F_i^{(L-p)} \otimes F_j^p \otimes \mathcal{O}_X(1).$$

and let $V = V_{i,j}$. We claim that for N sufficiently large, $V^N = H^0(X,M^{\otimes N})$. For V determines a very ample linear system and hence an embedding $X \subseteq P^m$, where $m+1 = \dim V$. There is a short exact sequence

$$0 \to \mathcal{J}_X(N) \to \mathcal{O}_{P^m}(N) \to \mathcal{O}_X(N) \to 0$$

For $N \gg 0$, $H^1(\mathcal{J}_X(N)) = 0$, so

$$S^N(V) = H^0(P^m,\ \mathcal{O}(N)) \twoheadrightarrow H^0(X,M^{\otimes N}).$$

Thus our claim follows. Releasing i,j,L and p, we have

$$V_{i,j}^N = H^0(X,(F_i^{\otimes(L-p)} \otimes F_j^p \otimes \mathcal{O}_X(1))^{\otimes N}) \tag{4.3.1}$$

By Riemann-Roch, the right hand side of (4.3.1) has dimension $N((L-p)e_i + pe_j + d) + 1 - g$. Note that $V_{i,j}^N$ is generated by monomials whose λ-weight is less than or equal to $((L-p)r_i + pr_j + r_\ell)N$. Consider the following filtration:

$$\begin{aligned}
&(V_{i_1}^L \cdot V_\ell)^N \subseteq (V_{i_1}^{L-1} \cdot V_{i_2} \cdot V_\ell)^N \subseteq \cdots \subseteq (V_{i_1} \cdot V_{i_2}^{L-1} \cdot V_\ell)^N \subseteq \\
&(V_{i_2}^L \cdot V_\ell)^N \subseteq (V_{i_2}^{L-1} \cdot V_{i_3} \cdot V_\ell)^N \subseteq \cdots \subseteq (V_{i_2} \cdot V_{i_3}^{L-1} \cdot V_\ell)^N \subseteq \\
&\qquad\vdots \\
&(V_{i_{k-1}}^L \cdot V_\ell)^N \subseteq \cdots \qquad \subseteq (V_{i_{k-1}} \cdot V_{i_k}^{L-1} \cdot V_\ell)^N \subseteq (V_{i_k}^L \cdot V_\ell)^N
\end{aligned}$$

We pick a basis of $(V_{i_k}^L \cdot V_\ell)^N = H^0(X, \mathcal{O}((L+1)N))$ by picking a basis of $(V_{i_1}^L \cdot V_\ell)^N$, then extending this to a basis of $(V_{i_1}^{L-1} \cdot V_{i_2} \cdot V_\ell)^N$, etc. We estimate the total weight T of such a basis

$$\begin{aligned}
T \le\ & N(Lr_{i_1} + r_\ell)Ne_\ell + \\
&+ \sum_{p=1}^{L} N((L-p)r_{i_1} + pr_{i_2} + r_\ell)(e_{i_2} - e_{i_1})N \\
&\quad\vdots \\
&+ \sum_{p=1}^{L} N((L-p)r_{i_{k-1}} + pr_{i_k} + r_\ell)(e_{i_\ell} - e_{i_{\ell-1}})N.
\end{aligned}$$

Thus

$$T \le N^2L^2\left\{\sum_f \frac{e_{i_{f+1}} - e_{i_f}}{2}(r_{i_{f+1}} + r_{i_f})\right\} + o(N^2, L^2)$$

where $o(N^2, L^2)$ denotes terms which are much smaller that N^2L^2 if $N \gg L \gg 0$. On the other hand, one sees that the term in the braces is

$$\left\{ r_\ell e_\ell - \sum_f \frac{e_{i_f} + e_{i_{f+1}}}{2}\left(r_{i_{f+1}} - r_{i_f}\right)\right\}$$

Thus Proposition 4.3 is established.

To deduce Theorem 4.2 from Proposition 4.3 we use the Riemann Roch Theorem and Clifford's Theorem to estimate e_k. If $e_k \ge 2g-1$, $e_k + 1 - g \ge k$, since $H^1(F_k) = 0$ and $H^0(F_k)$ has at least k sections. If $e_k \le 2g-2$, Clifford's Theorem says that $e_k \ge 2(k-1)$. If $d > 2g$, we see that in either case

$$e_k \ge \frac{d}{d-g}(k-1)$$

with strict inequality except for $k = 1$ and $k = \ell$. Given a sequence $1 = i_1 < \cdots < i_k = \ell$, to show

$$\sum \frac{e_{i_f} + e_{i_{f+1}}}{2}\left(r_{i_{f+1}} - r_{i_f}\right) > r_\ell d,$$

it suffices to show that

$$(4.3.2) \qquad \frac{d}{d-g} \sum_f \frac{i_f + i_{f+1} - 2}{2}\left(r_{i_{f+1}} - r_{i_f}\right) \geq r_\ell d$$

(The r_i's are assumed non-constant.) Consider the Newton polygon of the points (k,r_k). Let the i_f be the break points of this polygon, and let ρ_k be the point on the polygon above k. Then $\rho_k \leq r_k$, so $\sum \rho_k \leq 0$. To show (4.3.2) holds, it suffices to show that

$$(4.3.3) \qquad \frac{d}{d-g} \sum \frac{i_f + i_{f+1} - 2}{2}\left(\rho_{i_{f+1}} - \rho_{i_f}\right)$$

is $\geq \rho_\ell d$. Since the ρ_k are linear functions of k between i_f and i_{f+1}, (4.3.3) can be replaced by

$$(4.3.4) \qquad \frac{d}{d-g} \sum_{k=1}^{\ell-1} \frac{2k-1}{2} (\rho_{k+1} - \rho_k)$$

But (4.3.4) is just

$$\frac{d}{d-g}\left((\ell-1)\rho_\ell + \frac{1}{2}(\rho_1 + \rho_\ell) - \sum \rho_k\right)$$

Note $\sum \rho_k \leq 0$ and since the ρ's are convex functions of k, we have $\frac{\ell}{2}(\rho_1 + \rho_\ell) \geq \sum \rho_i$. Finally, $\ell - 1 = d - g$, so (4.3.2) is valid.

§5. Let us fix d and g and let $P(n) = nd + 1 - g$ and let U be the set of all subschemes of $\mathbb{P}(W)$ with Hilbert polynomial P. We defined the Hilbert point mapping

$$H_m : U \to \mathbb{P}(\overset{P(m)}{\Lambda} S^m(W))$$

which is injective. According to Grothendieck, $H_m(U)$ is the set of (closed) points of a closed subscheme $\mathcal{H} \subseteq \mathbb{P}(\overset{P(m)}{\Lambda} S^m(W))$. We are going to investigate the properties of a curve C so that $H_m(C)$ is semi-stable for $m >> 0$ and C is connected.

We assume $g > 1$ and $d \geq 1000g^2$. Our aim is to establish the following:

Proposition 5.1.

There is an M so that if $m \geq M$ and C is any curve of degree d and genus g which is m-Hilbert stable in $\mathbb{P}(W)$, then C is reduced, has only ordinary nodes as singularities, and ω_C has non-negative degree on each component of C. Further, any chain of rational curves on which ω_C is trivial has degree 1, i.e. consists of a straight line.

By being more careful, one can show the above holds in $d \geq 10(g-1)$. $[G_3, M_2]$.

The method of proof of Proposition 5.1 is to exhibit a 1-P-S λ of $SL(W)$ for each C which does not satisfy the criteria of the Proposition. We begin with some general definitions. Let $\mathfrak{F}$ be a coherent sheaf on a scheme and let $W \subseteq H^0(X,\mathfrak{F})$ be a subspace so that $\mathfrak{F}$ is generated at each point by the sections in W.

Definition 5.2

A weighted filtration on $\mathfrak{F}$

$$B = \begin{pmatrix} \mathfrak{F}_1 & \cdots & \mathfrak{F}_k \\ & & \\ r_1 & \cdots & r_k \end{pmatrix}$$

is a sequence of subsheaves $\mathfrak{F}_1 \subseteq \mathfrak{F}_2 \subseteq \cdots \subseteq \mathfrak{F}_k = \mathfrak{F}$ and rational numbers r_i, $r_1 \leq r_2 \leq \cdots \leq r_k$. Let $B' = \begin{pmatrix} \mathfrak{F}'_i \\ r'_i \end{pmatrix}$ be another weighted filtration of $\mathfrak{F}$. If $\mathfrak{F}_i \subseteq \mathfrak{F}'_j$ whenever $r_i \leq r'_j$, we say B' dominates B.

Let $\pi : Y \to X$. Given a weighted filtration $B = \begin{pmatrix} \mathcal{G}_i \\ r_i \end{pmatrix}$ on $\pi^*(\mathfrak{F})$, there is an induced filtration $B' = \begin{pmatrix} W_i \\ r_i \end{pmatrix}$ on W, where

$$W_i = \{s \in W \mid \pi^*(s) \in H^0(Y, \mathcal{G}_i)\}.$$

Conversely, given a weighted filtration $\begin{pmatrix} W_i \\ r_i \end{pmatrix}$ on W, there is an induced filtration on $\mathfrak{F}$. (These two processes of induced filtration do not commute).

Let L be a line bundle on a curve C and suppose $V \subseteq H^0(C,L)$ is a very ample linear system. Let $\begin{pmatrix} V_i \\ r_i \end{pmatrix} = B$ be a weighted filtration on V. We choose a basis (X_j,ρ_j) of V compatible with $\begin{pmatrix} V_i \\ r_i \end{pmatrix}$. (Thus $X_j \in V_i$ if $\rho_j \leq r_i$). We let $w_B(m,C)$ be the minimum weight of a basis of $H^0(C,L^{\otimes m})$ consisting in monomials in the X_i's . (One can show $w_B(m,C)$ is a polynomial of degree 2 in m if $m >> 0$) If $f(m)$ is a function of m, we write

$$w_B(m,C) \geq f(m) + O(m)$$

to mean that there is a constant K depending only on d and g so that

$$w_B(m,C) \geq f(m) + Km.$$

We now give two lemmas which will enable us to estimate $w_B(m,C)$ from below.

<u>Lemma 5.3</u>.

Suppose $C_i \subseteq C$ are subcurves. Suppose the natural map:

$$\varphi : \mathcal{O}_C \to \oplus\, \mathcal{O}_{C_i}$$

has kernel and cokernel of finite length. Then

$$w_B(m,C) \geq \sum_i w_B(m,C_i) + O(m).$$

<u>Proof</u>.

Let q be the maximum of the lengths of ker φ and coker φ. Then for $m >> 0$, the kernel and cokernel of

$$\varphi_m : H^0(C,L^{\otimes m}) \to \oplus\, H^0(C_i,L^{\otimes m})$$

have dimension $\leq q$. Given a basis $P_1,\ldots,P_r$ of $H^0(C,L^{\otimes m})$, we can by suitably reordering the P_i and partitioning $P_1,\ldots,P_{r-q}$ into sets $Q_i \subseteq \{P_1,\ldots,P_{r-q}\}$, assume that Q_i gives an independent set in $H^0(C_i,L)$. Thus

$$w_B(m,C) \geq \sum_i w_B(m,C_i) + m(r_1 - r_\ell) \cdot q.$$

For the next lemma, assume that C is irreducible. Let $\tilde{C}$ be the normalization of C_{red} and let $\mathcal{J}$ be the ideal of C_{red} in C. Let ℓ be the length of the local ring of the generic point of C. Suppose R is an effective divisor on $\tilde{C}$. Let p be an integer and suppose the $r_1,\ldots,r_k$ are integers.

Proposition 5.4.

Suppose V_j maps to zero in $H^0(\tilde{C},\tilde{L})$ for $j < p$ and that V_i maps to $H^0(\tilde{C},\tilde{L}(r_i - r_k)R))$. Then if $\deg L \geq (r_k - r_p) \deg R$, we have

$$w_B(m,C) \geq \frac{1}{2}((r_k - r_p)^2 \deg R + 2\ell r_p \deg \tilde{L})m^2 + O(m)$$

Proof.

First replace C by the subscheme defined by $\mathcal{J}^\ell$. Since $\mathcal{J}^\ell$ is supported at a finite number of points, neither the hypothesis nor the conclusion of the theorem are changed.

Let B' be the weighted filtration

$$\begin{pmatrix} V_p & \cdots & V_k \\ r_p & \cdots & r_k \end{pmatrix},$$

that is, we change the weights of the V_i for $i \leq p$ from r_i to r_p. Let (X_i,ρ_i) be a basis of V compatible with B. Let M be a monomial in the X_i's which is nonzero in $H^0(C,L^{\otimes m})$. Then M can involve at most ℓ of the X_i's with $X_i \in V_{p-1}$, since $\mathcal{J}^\ell = 0$. Thus

$$w_B(m,C) \geq w_{B'}(m,C) + O(m),$$

since the B and B' weight of a monomial can differ at most by $\ell(r_p - r_1)$.

Thus we may assume $B = B'$. Notice that

$$h^0(C,L^{\otimes m}) = m\ell \deg_{\tilde{C}}(\tilde{L}) + O(1).$$

Consider the new weighted filtration

$$B' = \begin{pmatrix} V_i \\ r_i - r_p \end{pmatrix}, \quad i \geq p.$$

We have

$$w_B(m,C) = w_{B'}(m,C) + mr_p h^0(C,L^{\otimes m}) = w_{B'}(m,C) + m^2 \ell r_p \deg \tilde{L} + O(m).$$

Hence it suffices to establish our Proposition for B'. So we may assume $r_p = 0$.

Since $r_i \geq 0$,

$$w_B(m,C) \geq w_B(m,C_{red}),$$

so we may assume C is reduced. Now let M be a monomial in $V^{\otimes m}$ of weight Q. The image of M is in $H^0(\tilde{C},\tilde{L}^{\otimes m}((Q-r_k m)R))$. Thus there is a constant C_1 so that the image of any monomial of weight Q or less lies in a subspace of codimension at least $(r_k m - Q)\deg R + C_1$ in $H^0(C,L^{\otimes m})$. Adding up the possible contributions for each weight Q, we see any basis must have weight at least

$$\sum_{Q=0}^{mr_k} (Q\deg R + O(1)) = r_k^2(\deg R)\frac{m^2}{2} + O(m)$$

Next suppose C is m-Hilbert semi-stable in $\mathbb{P}(W)$ and let $\begin{pmatrix} W_i \\ r_i \end{pmatrix}$ be a weighted filtration of W and assume $r_i \in \mathbb{Z}$. We let (X_i, ρ_i) be a compatible basis of W and let $w_B = \sum \rho_i$ be the weight of the basis. We claim

$$w_B(m,C) \leq (m^2 d)\frac{w_B}{n} \qquad (n = \dim W)$$

if $h^1(L^{\otimes m}) = 0$. Indeed, assign to weight ρ_i' to X_i by

$$\rho_i' = n\rho_i - w_B$$

Letting B' be the weighted filtration dervived from the ρ_i', we see that

$$n\, w_B(m,C) = w_{B'}(m,C) + h^0(L^{\otimes m})w_B m .$$

$$h^0(L^{\otimes m}) = md + 1 - g < md$$

$$\sum \rho_i' = 0$$

We can define a 1 - P.S. λ by $X_i^{\lambda(\alpha)} = \alpha^{\rho_i'} X_i$. Proposition 4.1 shows that $H_m(C)$ is unstable if $w_B(m,C) \geq \frac{1}{n}(m^2 d w_B)$. We will let

$$\frac{d}{n} = 1 + \varepsilon.$$

Since $n = d + 1 - g$, we see $\varepsilon = \frac{g-1}{n} \leq \frac{1}{1000g}$. <u>From now on, we assume there are no destabilizing</u> 1 - PS.

<u>Lemma 5.5</u>.

The components of C are generically reduced.

Proof.

Let $C_1 \subseteq C$ be a reduced and irreducible subcurve of C and let $\mathcal{J}$ be the ideal sheaf of C_1. Consider the filtration B induced on W by

$$\begin{pmatrix} L \otimes \mathcal{J} & L \\ 0 & 1 \end{pmatrix}$$

Let x be a generic point of C_1 and let ℓ be the length of $\mathcal{O}_{C,x}$. We can find two subcurves C_2 and C_3 in C so that $(C_2)_{red} = C_1$ and the natural map

$$\mathcal{O}_C \to \mathcal{O}_{C_2} \oplus \mathcal{O}_{C_3}$$

has kernel and cokernel of finite length.

Note that the filtration B is nontrivial. Let $d' = \deg_{C_1}(L)$. If B were trivial, W would map injectively to $H^0(C_1,L)$. But if $\ell \geq 2, d \geq 2d'$. Further $h^0(C_1,L) \geq n$ and $h^0(C_1,L) \leq d' + 1$, which is absurd. Note that $w_B \leq d' + 1$, so

$$w_B(m,C) \leq m^2 w_B(1 + \varepsilon) \leq (1 + \varepsilon)m^2(d' + 1)$$

On the other hand

$$w_B(m,C) \geq w_B(m,C_2) + w_B(m,C_3) + 0(m)$$

Further,

$$w_B(m,C_2) \geq \ell d'm^2 + 0(m)$$

from Proposition 5.4. Thus $\ell d' \leq d' + 1$ for $m \gg 0$, since $w_B(m,C_3) \geq 0$. Since $\ell \geq 2$, we see that $\ell = 2$, and $d' = 1$. In particular, $C_2 \cap C_3 \neq \emptyset$, since otherwise $C_3 = \emptyset$ and $d = 2$. Let C_4 be a component of C_3 meeting C_2. Let R be a point of $\tilde{C}_4$ mapping to C_1. Applying Proposition 5.4 we see that

$$w_B(C_4,m) \geq \frac{1}{2} m^2 + 0(m)$$

Hence

$$w_B(m,C) \geq (2d' + \frac{1}{2}) m^2 + 0(m)$$

So

$$2d' + 1/2 \leq d' + 1.$$

Lemma 5.6.

If $C' \subseteq C$ is a subcurve and $h^0(C'_{red}, L) \leq d'$, then $d' \geq 20\,g$.

Proof.

Look at the filtration on W induced by

$$\begin{pmatrix} L \otimes \mathcal{J}_{C'_{red}} & L \\ 0 & 1 \end{pmatrix}.$$

Then $w_B \leq d'$ by hypothesis. Further, if $d' < 20g$, $C' \neq C$. Let $C'' = \overline{C - C'}$. As in Lemma 5.5,

$$w_B(m, C'') \geq \frac{1}{2} m^2 + O(m)$$

$$w_B(m, C') \geq m^2 d' + O(m)$$

So

$$(d' + \tfrac{1}{2}) m^2 \leq d'(1 + \varepsilon) m^2 + O(m).$$

Since $\varepsilon \leq \frac{1}{1000\,g}$, our claim follows

Lemma 5.7.

If C' is a reduced irreducible subcurve of C, then the map $\pi : \tilde{C}' \to C$ is unramified.

Proof.

Note that if $\tilde{C}' \to C'$ is ramified, C' is a singular curve, so Lemma 5.6 is applicable. Let $R \in \tilde{C}'$ be a point of ramification and $\tilde{L} = \pi^*(L)$. Consider the filtration B induced on W by

$$\begin{pmatrix} \tilde{L}(-3R) & \tilde{L}(-2R) & \tilde{L} \\ 0 & 1 & 3 \end{pmatrix}$$

$$B = \begin{pmatrix} W_1 & W_2 & W_3 \\ 0 & 1 & 3 \end{pmatrix}$$

Since π is ramified at R, $\dim W_3/W_2 \leq 1$ and of course $\dim W_2/W_1 \leq 1$. So $w_B \leq 4$. From Proposition 5.4

$$w_B(m,C') \geq \frac{9}{2} m^2 + O(m) .$$

We reach a contradiction.

Lemma 5.8.

C_{red} has no triple points.

Proof.

Suppose first that three distinct components C_1, C_2 and C_3 meet in a point P. Let B be the weighted filtration induced on W by

$$\begin{pmatrix} L(-P) & L \\ 0 & 1 \end{pmatrix}$$

Now $w_B = 1$. As in the preceding lemmas,

$$w_B(m,C_i) \geq \frac{1}{2} m^2 + O(m)$$

So

$$w_B(m,C) \geq \frac{3}{2} m^2 + O(m)$$

But

$$w_B(m,C) \leq (1 + \varepsilon)m^2 .$$

Next, suppose that C_1 and C_2 meet in a singular point of C_1. Then C_1 is singular and so $\deg_{C_1}(L) \geq 20g$. Applying Proposition 5.4, we see that

$$w_B(m,C_1) \geq m^2 + O(m) .$$

So we again obtain a contradiction.

Similarly, C_1 cannot have a triple point.

Lemma 5.8.

C_{red} has no tacnodes.

Proof.

Suppose C_1 and C_2 meet at P and that their tangent lines are identical. Let d_i be the degree of L on C_i. We may suppose $d_1 \leq d_2$. Consider the filtration B_i on W induced by

$$\begin{pmatrix} \tilde{L}_i(-2P) & \tilde{L}_i(-P) & \tilde{L}_i \\ 0 & 1 & 2 \end{pmatrix}$$

Our assumption implies that $B_1 = B_2$ and so $w_{B_1} \leq 3$.

If $d_i \geq 2$, we see that

$$w_{B_1}(m,C_i) \geq 2m^2 + O(m)$$

If $d_i = 1$,

$$w_{B_1}(m,C_i) \geq \frac{1}{2}(1 + 2)m^2 = \frac{3}{2}m^2 + O(m).$$

Since we cannot have $d_1 = d_2 = 1$, we see that

$$w_{B_1}(m,C) \geq \frac{7}{2}m^2 + O(m).$$

As before, we reach a contradiction.

Putting together our results so far, we see that C_{red} has only nodes as singularities. Next, we want to show that $H^1(C_{red},L) = 0$. We begin with a version of Clifford's Theorem:

Lemma 5.9.

Let C be a reduced curve with only nodes and let L be a line bundle on C generated by global sections. If $H^1(C,L) \neq 0$, there is a curve $C' \subseteq C$ so that

$$h^0(C',L) \leq \frac{\deg_{C'} L}{2} + 1.$$

This is Lemma 9.1 of [G-M].

Suppose $H^1(C_{red},L) \neq 0$ and let C' be the curve of the above lemma and let $d' = \deg_{C'} L$. Consider the filtration induced on W by

$$\begin{pmatrix} L \otimes \mathcal{I}_{C'_{red}} & L \\ 0 & 1 \end{pmatrix}$$

Then $w_B \leq \frac{d'}{2} + 1$. On the other hand, Proposition 5.4 shows that

$$w_B(m,C) \geq d'm^2 + O(m) .$$

As usual, we reach a contradiction unless $d' \leq 2$. Thus C' is either P^1 or two copies of P^1 meeting at one point. But then we do not have

$$h^0(C',L) \leq \frac{d'}{2} + 1.$$

Proposition 5.10

C is reduced and $W = H^0(C,L)$.

Proof.

Consider $\mathcal{J}$, the ideal defining C_{red} in C. $\mathcal{J}$ is supported at a finite number of points. We claim

$$(5.10.1) \qquad W \cap H^0(C,\mathcal{J}.L) \neq 0$$

if $\mathcal{J} \neq 0$. Suppose not. Let g' be the genus of C_{red} and let ℓ be the length of $\mathcal{J}$. We have

$$g' = g + \ell.$$

Thus if $\ell > 0$,

$$h^0(C_{red},L) < \deg L + (1 - g) = \dim W.$$

since $H^1(C_{red},L) = 0$. Thus (5.10.1) is established.

Consider the filtration B on W induced by

$$\begin{pmatrix} L \otimes \mathcal{J} & L \\ 0 & 1 \end{pmatrix} .$$

Then $w_B < \dim W$, but

$$w_B(m,C) \geq m^2 \dim W + O(m).$$

Once again we reach a contradition.

Next let $C' \subseteq C$ be a subcurve and let ω_C be the dualizing sheaf. Let e be $\deg_{C'}L$ and suppose k is the number of points in $C' \cap \overline{C - C'}$.

Proposition 5.11.

$$(1+\varepsilon)h^0(C',L) \geq e + k/2 .$$

Proof.

Suppose not. Let $\mathcal{I}_{C'}$ be the ideal of C' and consider the filtration B induced on W by

$$\begin{pmatrix} \mathcal{I}_{C'}.L & L \\ 0 & 1 \end{pmatrix}$$

Then $w_B \leq h^0(C',L)$.

Now

$$w_B(m,C') \geq em^2 + 0(m)$$

For the moment, assume that for every irreducible component C_j of $\overline{C - C'}$, we have

(5.11.1) $$2\deg_{C_j}(L) \geq \#(C_j \cap C')$$

On the one hand, if $\deg_{C_j}(L) \geq \#(C_j \cap C')$, we have

$$w_B(m,C_j) \geq \frac{1}{2}(\#(C_j \cap C'))m^2 + 0(m)$$

On the other hand, if $\deg_{C_j}(L) < \#(C_j \cap C')$, then every section of $\mathcal{I}_{C'}.L$ vanishes on C_j, so

$$w_B(m,C_j) \geq \deg_{C_j}(L)m^2 + 0(m)$$

$$\geq \frac{1}{2}(\#(C_j \cap C')) + 0(m)$$

So in either case, we have

$$w_B(m,C_j) \geq \frac{1}{2}(\#(C_j \cap C')) .$$

Hence in the presence of our hypothesis (5.11.1) we see that

$$w_B(m,C) \geq (e + \frac{1}{2}\sum_j(\#(C_j \cap C))) m^2 + 0(m) \geq (e + \frac{1}{2}k)m^2 + 0(m) .$$

So the Proposition is established if (5.11.1) holds.

Suppose C' is irreducible and that $e < k/2$ and let C_j be a component of $\overline{C - C'}$. We claim

$$(5.11.2) \qquad \deg_{C_j}(L) \geq \frac{1}{2}(C_j \cap C')$$

This is certainly true if $\#(C_j \cap C') = 1$. So assume $\#(C_j \cap C') \geq 2$. Assume $\deg_{C_j}(L) < \frac{1}{2}(C_j \cap C')$. Then the genus of $C_j \cup C'$ is positive, so $C + \deg_{C_j}(L) \geq 20g$ by Lemma 5.6. $\#(C_j \cap C')$ is not greater that the genus of $C_j \cup C'$, so

$$\#(C_j \cap C') \leq g.$$

We have reached a contradiction, so (5.11.2) is established.

We can now conclude from the first part of the proof that

$$(1 + \varepsilon)(e + 1) \geq (1 + \varepsilon)h^0(C',L) \geq e + \frac{k}{2}$$

This contradicts $e < k/2$, since $k \leq g$. So $e > k/2$. So Proposition 5.11 is established.

Note that C is semi-stable as a curve, meaning that it has no nonsingular rational component C'' meeting the rest of the curve in exactly one point. Indeed, one just applies Proposition 5.11 to the curve $\overline{C - C''}$. One also sees that any chain of rational curves meeting the rest of C in two points must have length one and degree 1 by applying Proposition 5.11 to $\overline{C - C''}$. Also note that since $L_{C'}$ is a quotient of L, we must have $H^1(L_{C'}) = 0$, so Proposition 5.11 may be restated as

$$(1 + \varepsilon)(e + 1 - g') \geq e + k/2$$

where g' is the genus of C'.

Definition 5.12.

A line bundle L on a sėmi-stable curve X of genus g is said to be potentially stable if it has positive degree on all components and if $X' \subseteq X$ is any subcurve of genus g' and L degree e, we have

$$(1 + \varepsilon)(e + 1 - g') \geq e + k/2.$$

We have established that all m-Hilbert stable curves are potentially stable.

§6.

Proposition 6.1.

Suppose C is a stable curve and that $\mathcal{O}(1) \cong \omega_C^{\otimes \nu}$ for some $\nu \geq 1000g^2$. Then C is m-Hilbert stable for $m \gg 0$.

Proof.

We can find a smooth curve S, a point $P \in S$, and a flat family of curves of genus g $\pi : X \to S$ so that $X_P = C$ and so that X_Q is smooth for $Q \neq P$. We choose a basis for $\pi_*(\omega_{X/S}^{\otimes n})$ over $U - P$, where U is some neighborhood of P. Such a basis determines a map Φ of $U - P$ into m-Hilbert stable points of the Hilbert scheme $\mathcal{H}_{S.S.}$. Let $m : \mathcal{H}_{S.S.} \to \mathbb{P}^k$ be the invariant map of Theorem 1.2. Then using Corollary 1.2.1, we can assume by replacing $C - P$ by some ramified cover that Φ extends to a morphism of U to $\mathcal{H}_{S.S.}$ and that $\Phi(P)$ has infinite stabilizer if $\Phi(P)$ is not stable. Thus there is a family $p : Y \to S$ and a line bundle L on Y so that (Y,L) is isomorphic to $(X,\omega^{\otimes n})$ over $S - P$, and so that $H^1(L_P) = 0$ and the embedding of Y_P determined by the sections of L_P is very ample and is m-Hilbert stable for some fixed large m. The singularities of Y are given locally by $\{xy = t^k\}$ as they are deformations of simple nodes. By resolving these singularities we can obtain a new smooth surface $p' : Y' \to S$ so that Y'_P is semi-stable as a curve, since the resolution of $\{xy = t^n\}$ is just a chain of $(n-1)$ rational curves.

Denote the pullback of L to Y' by L again. There is an isomorphism of $\omega^{\otimes n}$ with L defined on Y' over $S - P$. It follows that we can write $\omega^{\otimes n} = L(-n_i E_i)$ on Y', where the E_i are components of Y'_P. We may assume $\min\{n_i\} = 0$. Let C' be a connected subcurve so that $n_i = 0$ if $E_i \subseteq C'$, but $n_i > 0$ if $E_i \not\subseteq C'$ and $E_i \cap C' \neq \emptyset$. Both $\omega^{\otimes n}$ and L are trivial on the new rational chains introduced by resolution of the singularities of Y, so C' either contains such a chain or meets it at one point. So C' is the proper transform of a subcurve of Y_P. Let e be the degree L on C', let g' be the genus of C' and let $k = \#C' \cap \overline{(Y'_P - C')}$. We have

$$e \leq \deg_{C'}\omega^{\otimes \nu} - k \tag{6.1.1}$$

$$(1 + \varepsilon)(e + 1 - g') \geq e + k/2 \tag{6.1.2}$$

$$\deg_{C'}\omega = \deg_{C'}\omega' + k = (2g' - 2) + k. \tag{6.1.3}$$

$$1 + \varepsilon = \frac{d}{n} = \frac{2\nu}{2\nu - 1} \tag{6.1.4}$$

Substituting (6.1.3) and (6.1.4) into (6.1.2), we obtain

$$\frac{2\nu}{2\nu - 1}\left(e - \frac{1}{2}(\deg_{C'}(\omega) - k)\right) \geq e + k/2$$

Simplifying, we get

$$e - \nu \deg_{C'} \omega \geq k/2 .$$

Thus we contradict (6.1.1.) unless $C' = Y_P'$, i.e., $\omega^{\otimes\nu} \cong L$.

One can use Proposition 6.1 to produce a projective moduli space for stable curves. See $[G_3, M_1]$ for details.

§7. Let X_o be an automorphism free irreducible stable curve of genus $g > 1$ with exactly one node N and let d be a large integer. Using techniques similar to those used in the proof of Theorem 6.1, one can show that if L is a line bundle of degree d, then the image of X_o under the linear system $W = H^0(X_o, L)$ is m-Hilbert stable for $m >> 0$.

Now let U_{X_o} be the subscheme of the Hilbert scheme $\mathcal{H}$ of P^{n-1} consisting of m-Hilbert stable curves of degree d which are isomorphic to X_o. $SL(n)$ operates on $\mathcal{H}$ and hence on U_{X_o}. Since we have assumed that X_o is automorphism free, the set of points of the quotient of U_{X_o} by $SL(W)$ can be identified with the generalized Jacobian $J^d_{X_o}$ of X_o. Indeed, every curve isomorphic to X_o in P^{n-1} determines a line bundle on X_o, namely $\mathcal{O}_{X_o}(1)$. So there is a map from U_{X_o} to J_{X_o}. One can show this map induces an isomorphism of U_{X_o}/G with $J^d_{X_o}$. Let V_{X_o} be $\bar{U}_{X_o} \cap \mathcal{H}_{S.S.}$. So a point of V_{X_o} corresponds to an m-Hilbert semi-stable curve which is a limit of curves in U_{X_o}. V_{X_o} is closed in $\mathcal{H}_{S.S.}$, so V_{X_o}/G is projective. Thus V_{X_o}/G is a projective compactification of $J^d_{X_o}$. Our aim is explicate the nature of this compactification.

First of all, we introduce the semi-stable models of X_o. Let X be the normalization of X_o and let $P_1, P_2 \in X$ be the two points mapping to the node of X_o. Define a series of semi-stable curves X_k for $k \geq 0$ by letting X_k be the curves whose components are X and non-singular rational curves $R_1, \ldots, R_k$ arranged as follows:

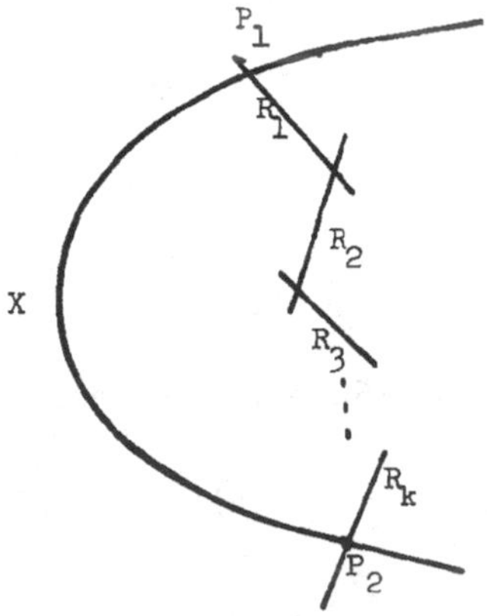

Fig. 1

Suppose S is a smooth curve and $p_1 : Y \to S$ is a proper flat family of curves parameterized by S. Suppose that Y_s is isomorphic to X_o for $s \neq s_o$. One can show that Y_{s_o} must be isomorphic to X_k for some k.

Thus the underlying curve C of an point of V_{X_o} must be X_k for some k. We have already noted that $k \leq 1$ and if $k = 1$, that the degree of the one rational curve is one. We claim that no non-trivial element of $SL(n)$ fixes C as a set. (Note: X_1 has automorphism group k^* as an abstract curve, but none of these automorphisms are induced by projective linear transformations.) Indeed any projective transformation σ which fixes $C \cong X_1$ must fix X and R_1 and hence P_1 and P_2. Since we assume X_o automorphism free, σ is the identity on X. But X does not lie in a hyperplane, since $X \cup R_1$ does not lie in a hyperplane and R_1 is a straight line joining two points of X. Since σ fixes X, σ fixes all of projective space. It follows that all the curves in V_{X_o} are actually stable, since any invariant set of strictly semi-staple curves has semi-stable curves with infinite stablilzer in its closure.

There is a simple construction which illustrates how the points of V_{X_o} are limits of points in U_{X_o}. Let S be a smooth curve and let $R \in S$ be fixed. Consider the surface $S \times X$. $S \times X$ has two natural sections, $S \times P_1$ and $S \times P_2$. If E is a bundle on $S \times X$, we let E_{P_i} be the result of restricting E to $S \times P_i$. Usually, we regard E_{P_i} as a bundle on S. Given an isomorphism $\varphi : E_{P_1} \to E_{P_2}$ over S, we can define a bundle on $X_0 \times S$ by using φ as descent data. Suppose that E is a line bundle L of degree d on $X \times S$ but that $\varphi : L_{P_1} \to L_{P_2}$ is only an isomorphism over $U = S - R$. Then φ determines a family of line bundles on X_o parameterized by U and hence a map ψ of U to U_{X_o}/G. We know that one can extend ψ to a map of S to V_{X_o}/G. Our aim is to see explictly how such an extension can be obtained for certain φ.

We assume first that φ is an isomorphism of L_{P_1} with $L_{P_2}(-R)$. In this case, we blow up the point $P_2 \times R$ in $X \times S$ to obtain a new surface $\mathfrak{X}_1$:

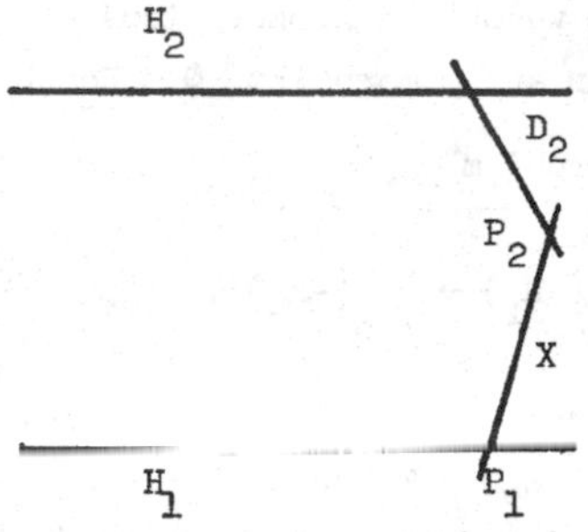

Fig. 2

D_2 denotes the new exceptional divisor, H_1 and H_2 the proper transforms of $P_i \times S$, and X the proper transform of $X \times R$. Let L again denote the pullback of L to X_1 and consider $L' = L \otimes \mathcal{J}_{D_2}$ where $\mathcal{J}_{D_2}$ is the ideal sheaf of D_2. Note $\deg_{D_2} L' = 1$, since $D_2^2 = -1$. Further, $L'_{H_2} = L_{P_2}(-R)$ and $L'_{H_1} = L_{P_1}$. Hence φ actually gives an isomorphism of L'_{H_1} with L'_{H_2}. We form a new family of curves $p: Z \to S$ by gluing H_1 to H_2 and then use φ to glue L'_{H_1} to L'_{H_2} to obtain a line bundle $\mathcal{L}$ on Z. We call $(Z,\mathcal{L})$ the geometric realization of φ. Note that Z_R is just X_1 and $\mathcal{L}_R$ has degree one on $R_1 \subseteq X_1$ and degree $d-1$ on X. One can show that the embedding of X_1 by the linear system $H^0(X_1,\mathcal{L}_R)$ is m-Hilbert stable for $m \gg 0$ by using an argument similar to that used to establish Proposition 6.1. Note the $(\mathcal{L}_R)_X$ is just $L_R(-P_2)$.

Define a map ψ from $V_{X_o} - U_{X_o}$ to J_X^{d-1} by sending a m-Hilbert stable curve $X_1 \subseteq \mathbf{P}^{n-1}$ to the line bundle $\mathcal{O}(1)_X$. ψ is onto, since given a line bundle M of degree $d-1$ on X, we let L on $X \times S$ be the constant bundle $M(+P_2)$ and choose an isomorphism φ of L_{P_1} with $L_{P_2}(-R)$. Then the geometric realization of φ yields a line bundle on X_1 which is isomorphic to M on X. On the other hand, the line bundle $\mathcal{O}(1)_X$ determines $X_1 \subseteq \mathbf{P}^{n-1}$ up to projective equivelence. Indeed, $\mathcal{O}_X(1)$ determines the map of X to $\mathbf{P}^{n-1}$ up to projective equivelence. But given $X \subseteq \mathbf{P}^{n-1}$, X_1 is determined since X_1 is just the union of X and the straight line joining P_1 and P_2. So we see that $V_{X_o} - U_{X_o}/SL(W)$ is just J_X^{d-1}, at least set theoretically.

We next give a more global version of geometric realization. This time start with $S_o = J_X^{d-1}$. On $S_o \times X$ we have the Poincaré bundle $\mathcal{L}$. Consider $S_1 = \mathbf{P}(\underset{\sim}{\mathrm{Hom}}(\mathcal{L}_{P_2},\mathcal{L}_{P_1}) \oplus \mathcal{O})$ and let M_1 be the tautological bundle for $\pi_1: S_1 \to S_0$

$$\mathcal{O} \oplus \underset{\sim}{\mathrm{Hom}}(\mathcal{L}_{P_2},\mathcal{L}_{P_1}) \to M_1 \to 0 .$$

There are disjoint divisors H_1 and H_2 of S_1 so that the map from $\mathcal{O}$ to M_1 vanishes on H_1 and the map from $\underset{\sim}{\mathrm{Hom}}(\mathcal{L}_{P_2},\mathcal{L}_{P_1})$ to M_2 vanishes on H_2. H_1 and H_2 are both sections of π_1. Thus we obtain isomorphisms $M_1 \cong \mathcal{O}(H_1)$ and $M_1 \cong \underset{\sim}{\mathrm{Hom}}(L_{P_2},L_{P_1})(H_2)$. So we obtain an isomorphism $\mathcal{O} \to \underset{\sim}{\mathrm{Hom}}(L_{P_2},L_{P_1})(H_2 - H_1)$. This can be interpreted as an isomorphism

$$\Theta: L_{P_1}(-H_1) \to L_{P_2}(-H_2) .$$

Now on $S_1 \times X$, blow up $H_1 \times P_1$ and $H_2 \times P_2$ to obtain new exceptional divisors E_1 and E_2. Consider $L' = L(-E_1 - E_2)$. We have $L'_{P_1'} \cong L_{P_1}(-H_1)$ and $L'_{P_2'} \cong L_{P_2}(-H_2)$, where the P_i' are the proper transtfons of $P_1 \times S_1$ and $P_2 \times S_2$.

We obtain a new family of semi-stable curves Y' by gluing P_1' to P_2' and using φ to descend L' to obtain a line bundle L' on Y'. Let $\pi' : Y' \to S_1$ be the projection. $\pi_*'(L')$ is locally free on S_1. Choosing a basis of $\pi_*'(L')$ over an open U determines a map of U to V_{X_o} and hence to $V_{X_o}/SL(n)$. This map is independent of the choice of the basis of $\pi_*'(L')$. We obtain a map ψ of S_1 to V_{X_0}/G. One can check that H_1 and H_2 both map isomorphically to $(V_{X_0} - U_{X_0})/SL(W)$. Both H_1 and H_2 are isomorphic to J_{d-1}. Let $\varphi : H_1 \to H_2$ be the map induced on J_{d-1} by sending a line bundle L to $L(P_2 - P_1)$. One can check that $\psi \circ \varphi = \psi$ on H_1. Although it is not obvious, one can show that $V_{X_0}/SL(W)$ is isomorphic to the normal crossing variety obtained by gluing H_1 to H_2 by φ. One first shows that V_{X_0} is a normal crossing variety by studying the deformations of X_1 and the uses the fact that $SL(W)$ is operating freely on V_{X_0} to show that $V_{X_0}/SL(W)$ has normal crossings. Thus $S_1 \to V_{X_0}/SL(W)$ is identified with the normalizion of $V_{X_0}/SL(W)$. Cf $[G_4]$.

References

[G_1] Gieseker, D., On the moduli of vector bundles on an algebraic surface. Ann. of Math 106, 45 (1977).

[G_2] __________, Global moduli for surfaces of general type. Inv. Math. 43, 233 (1977).

[G_3] __________, Lectures on Stable Curves, Tata Lecture Notes.

[G_4] __________, A degeneration of the moduli space of stable bundles. (To appear)

[G-M] __________ and I. Morrison, Hilbert stability of rank two bundles on curves.

[Ma] Maruyama, M., Moduli of stable sheaves I, II J. Math. Soc. Kyoto 17, 91 (1977) and 18, 557, (1978).

[M_1] Mumford, D. and J. Fogarty. Geometric Invariant Theory. Second Enlarged Edition. Springer Verlag 1982.

[M_2] __________, Stability of projective varieties. L'Ens Math. 24 (1977).

[N] Newstead, P. E. Introduction to Moduli Problems and Orbit Spaces. Tata Inst. Lecture Notes, Springer Verlag, 1978.

ROOT SYSTEMS, REPRESENTATIONS OF QUIVERS AND INVARIANT THEORY

Victor G. Kac

In these notes I will discuss two approaches to the study of the orbits, invariants, etc, of a linear reductive group G operating on a finite dimensional vector space V. The two techniques are the "quiver method" and the "slice method", which are discussed in Chapters I and II respectively.

Undoubtedly, the slice method, based on Luna's slice theorem [1], is one of the most powerful methods in geometric invariant theory. Even in the case of binary forms the slice method gives results which were out of reach of mathematucs of 19th century (cf. [15]). For example, I show that for the action of $SL_2(\mathbb{C})$ on the space of binary forms of odd degree $d > 3$ the minimal number of generators of the algebra of invariant polynomials is greater than $p(d-2)$, where $p(n)$ is the classical partition function.

On the other hand, the quiver method can be applied to a (very special) class of representations for which the slice method often fails.

Most of the results of Chapter I are contained in [4] and [5]; on the most part I just give simpler versions of the proofs. Chapter II contains some new results (as, it seems, the one mentioned above).

I am mostly greatful to the organizers of the summer school in Montecatini Terme (Italy) for inviting me to give these lectures, especially to F. Gherardelli who convinced me to write the notes. My thanks go to J. Dixmier for sharing his knowledge and enthusiasm about invariant theory of binary forms, and to H. Kraft and R. Stanley for several important observations.

Chapter I. Representations of quivers.

The whole range of problems of linear algebra can be formulated in a uniform way in the context of representations of quivers introduced by Gabriel [2]. In this chapter I discuss the links of this with invariant theory and theory of generalized root systems ([3], [4]).

§1.1. Given a connected graph Γ with n vertices $\{1,\ldots,n\}$ we introduce the associated root system $\Delta(\Gamma)$ as a subset in $\mathbb{Z}^n$ as follows. Let b_{ij} denote the number of edges connecting vertices i and j, if $i \neq j$, and twice the number of loops at i if $i=j$. Let $\alpha_i = (\delta_{i1},\ldots,\delta_{in})$, $i=1,\ldots,n$, be the standard basis of $\mathbb{Z}^n$. Introduce a bilinear form $(\ ,\ )$ on $\mathbb{Z}^n$ by:

$$(\alpha_i,\alpha_j) = \delta_{ij} - \tfrac{1}{2}b_{ij} \quad (i,j=1,\ldots,n).$$

Denote by $Q(\alpha)$ the associated quadratic form. It is clear that this is a $\mathbb{Z}$-valued form. The element α_i is called a fundamental root if there is no edges-loops at the vertex i. Denote by Π the set of fundamental roots. For a fundamental root α define the fundamental reflection $r_\alpha \in \operatorname{Aut}\mathbb{Z}^n$ by

$$r_\alpha(\lambda) = \lambda - 2(\lambda,\alpha)\alpha \quad \text{for} \quad \lambda \in \mathbb{Z}^n.$$

This is a reflection since $(\alpha,\alpha) = 1$ and hence $r_\alpha(\alpha) = -\alpha$, and also $r_\alpha(\lambda) = \lambda$ if $(\lambda,\alpha) = 0$. In particular, $(r_\alpha(\lambda), r_\alpha(\lambda)) = (\lambda,\lambda)$. The group $W(\Gamma) \subset \operatorname{Aut}\mathbb{Z}^n$ generated by all fundamental reflections is called the Weyl group of the graph Γ (for example $W = \{1\}$ if there is an edge-loop at any vertex of Γ). Note that the bilinear form $(\ ,\ )$ is $W(\Gamma)$-invariant. Define the set of real roots $\Delta^{re}(\Gamma)$ by:

$$\Delta^{re}(\Gamma) = \bigcup_{w\in W} w(\Pi).$$

For an element $\alpha = \sum_i k_i\alpha_i \in \mathbb{Z}^n$ we call the height of α (write: $\mathrm{ht}\,\alpha$) the number $\sum k_i$; we call the support of α (write: $\operatorname{supp}\alpha$) the subgraph of Γ consisting of those vertices i for which $k_i \neq 0$ and all the edges joining these vertices. Define the fundamental set $M \subset \mathbb{Z}^n$ by:

$$M = \{\alpha \in \mathbb{Z}_+^n\setminus\{0\} \mid (\alpha,\alpha_i) \le 0 \text{ for all } \alpha_i \in \Pi, \text{ and } \operatorname{supp}\alpha \text{ is connected}\}.$$

(Note that $(\alpha,\alpha_i) \le 0$ if $\alpha_i \notin \Pi$, automatically).

Here and further on, $\mathbb{Z}_+ = \{0,1,2,\ldots\}$.

Define the set of <u>imaginary roots</u> $\Delta^{im}(\Gamma)$ by:

$$\Delta^{im}(\Gamma) = \bigcup_{w \in W} w(M \cup -M).$$

Then the <u>root system</u> $\Delta(\Gamma)$ is defined as

$$\Delta(\Gamma) = \Delta^{re}(\Gamma) \cup \Delta^{im}(\Gamma).$$

An element $\alpha \in \Delta(\Gamma) \cap \mathbb{Z}_+^n$ is called a <u>positive root</u>. Denote by $\Delta_+(\Gamma)$ (resp. $\Delta_+^{re}(\Gamma)$ or $\Delta_+^{im}(\Gamma)$) the set of all positive (resp. positive real or positive imaginary) roots. When it does not cause a confusion we will write W, Δ, etc. instead of $W(\Gamma)$, $\Delta(\Gamma)$, etc.

It is obvious that $(\alpha,\alpha) = 1$ if $\alpha \in \Delta^{re}$. On the other hand, $(\alpha,\alpha) \leq 0$ if $\alpha \in \Delta^{im}$. (Indeed, one can assume that $\alpha = \Sigma k_i \alpha_i \in M$; but then $(\alpha,\alpha) = \sum_i k_i(\alpha,\alpha_i) \leq 0$.) Hence

$$\Delta^{re} \cap \Delta^{im} = \emptyset.$$

Furthermore, one has:

$$\Delta = \Delta_+ \bigsqcup -\Delta_+.$$

This statement is less obvious but will follow from the representation theory of quivers.

We shall need two more easy facts:

$$\Delta_+^{im} = \bigcup_{w \in W} w(M) = \{\alpha \in \Delta_+ \mid W(\alpha) \subset \Delta_+\};$$

$\Delta_+^{re} = \{\alpha \in \Delta_+ \setminus \{\alpha_1,\ldots,\alpha_n\} \mid$ there exist $\alpha_1,\ldots,\alpha_s \in \Pi$ such that $r_{\alpha_i} r_{\alpha_{i+1}} \cdots r_{\alpha_s}(\alpha) \in \Delta_+$ for $1 \leq i \leq s$ and $r_{\alpha_1} \cdots r_{\alpha_s}(\alpha) \in \Pi\} \cup \Pi$.

The proof of these facts can be found in [3].

§1.2. According as the bilinear form (,) is positive definite, positive semidefinite or indefinite the (connected) graph Γ is called a graph of <u>finite</u>, <u>tame</u> and <u>wild type</u> respectively. The complete lists of finite and tame graphs are given in Tables F and T.

<u>Table F.</u>

A_n $(n \geq 1)$

D_n $(n \geq 4)$

E_6

E_7

E_8

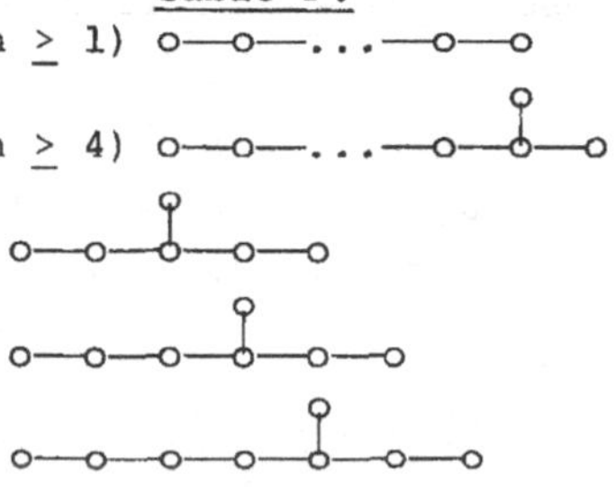

The subscript in the notation of a graph in Table F equals to the number of vertices.

Table T.

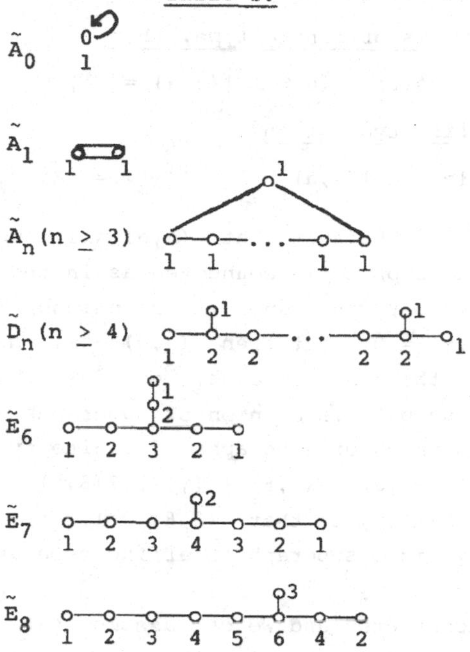

$\tilde{A}_0$

$\tilde{A}_1$

$\tilde{A}_n (n \geq 3)$

$\tilde{D}_n (n \geq 4)$

$\tilde{E}_6$

$\tilde{E}_7$

$\tilde{E}_8$

The subscript in the notation of a graph in Table T plus 1 equals to the number of vertices. The kernel of the bilinear form (,) is $\mathbb{Z}\delta$, where $\delta = \sum_i a_i \alpha_i$, a_i being the lables by the vertices. It is easy to show that the converse is also true (see e.g. [3], p.61):

Proposition. If there exists $\delta \in \mathbb{Z}_+^n$ such that $(\delta, \alpha_i) = 0$ for all i and $\delta \neq 0$ then Γ is of tame type.

Here are some characterisations of graphs of finite, tame and wild types:

Γ is finite $\iff$ $|W(\Gamma)| < \infty \iff |\Delta(\Gamma)| < \infty \iff \Delta^{im}(\Gamma) = \emptyset$,

Γ is tame $\iff$ $\Delta^{im}(\Gamma)$ lies on a line;

Γ is wild $\iff$ there exists $\alpha \in \Delta_+(\Gamma)$ such that $(\alpha, \alpha_i) < 0$ for all i and supp $\alpha = \Gamma$.

A graph of wild type is called hyperbolic if every one of its proper connected subgraph is of finite or tame type. In the case of a finite, affine or hyperbolic graph, there is a simple description of the root system $\Delta(\Gamma)$.

<u>Proposition</u>. <u>If a graph</u> Γ <u>is of finite, affine or hyperbolic type, then</u>

$$\Delta(\Gamma) = \{\alpha \in \mathbb{Z}^n\setminus\{0\} \mid (\alpha,\alpha) \leq 1\}.$$

<u>In particular, if</u> Γ <u>is of finite type, then</u>

$$\Delta(\Gamma) = \{\alpha \in \mathbb{Z}^n \mid (\alpha,\alpha) = 1\},$$

<u>and if</u> Γ <u>is of affine type, then</u>

$$\Delta^{re}(\Gamma) = \{\alpha \in \mathbb{Z}^n \mid (\alpha,\alpha) = 1\}; \Delta^{im}(\Gamma) = (\mathbb{Z}\setminus\{0\})\delta.$$

<u>Proof</u>. Let $\alpha \in \mathbb{Z}^n\setminus\{0\}$ be such that $(\alpha,\alpha) \leq 1$. We have to show that $\alpha \in \Delta(\Gamma)$. Note that $\operatorname{supp}\alpha$ is connected as in the contrary case $\alpha = \beta + \gamma$, where $\operatorname{supp}\beta$ and $\operatorname{supp}\gamma$ are unions of subgraphs of finite type and $(\beta,\gamma) = 0$, but then $(\alpha,\alpha) \geq 2$. Next, either α or $-\alpha \in \mathbb{Z}^n_+$. Indeed, in the contrary case, $\alpha = \beta-\gamma$, where $\beta,\gamma \in \mathbb{Z}^n_+$, $\operatorname{supp}\beta \cap \operatorname{supp}\gamma = \emptyset$, $\operatorname{supp}\beta$ is a union of subgraphs of finite type and $\operatorname{supp}\gamma$ is either a union of subgraphs of finite type or is a subgraph of affine type. But $(\alpha,\alpha) = (\beta,\beta) + (\gamma,\gamma) - 2(\beta,\gamma) \leq 1$ and $(\beta,\gamma) \leq 0$. Hence the only possiblility is that $(\beta,\beta) = 1$, $(\gamma,\gamma) = 0$ and $(\beta,\gamma) = 0$. But then $\operatorname{supp}\gamma$ is a subgraph of affine type and (β,γ) must be < 0, a contradiction.

So, $\operatorname{supp}\alpha$ is connected and we can assume that $\alpha \in \mathbb{Z}^n_+$. We can assume that $W(\alpha) \cap \Pi = \emptyset$, otherwise there is nothing to prove. But then, clearly, $W(\alpha) \in \mathbb{Z}^n_+$. Taking in $W(\alpha)$ an element of minimal height, we can assume that $(\alpha,\alpha_i) \leq 0$ for $\alpha_i \in \Pi$. Since, in addition, $\operatorname{supp}\alpha$ is connected, we deduce that α lies in the fundamental set. □

Using the proposition, one can describe $\Delta^{re}_+(\Gamma)$ for a tame graph $\Gamma = \tilde{A}_n, \tilde{D}_n$ or $\tilde{E}_n$ via the subset $\mathring{\Delta}_+ \subset \Delta^{re}_+(\Gamma)$ of positive roots of the subgraph A_n, D_n or E_n respectively as follows:

$$\Delta^{re}_+(\Gamma) = \{\alpha + n\delta \mid \pm\alpha \in \mathring{\Delta}_+, n \geq 1\} \cup \mathring{\Delta}_+.$$

One can show, that conversely, if $\Delta(\Gamma) = \{\alpha \in \mathbb{Z}^n \mid (\alpha,\alpha) \leq 1\}$, then Γ is of finite, affine or hyperbolic type.

<u>Remark</u>. The graphs of finite type are the so called simply laced Dynkin diagrams. They correspond to simple finite-dimensional Lie algebras with equal root length. The other graphs correspond to certain infinite-dimensional Lie algebras, the so called Kac-Moody algebras.

§1.3. Examples.

a) Denote by S_m the graph with one vertex and m edges loops. The associated quadratic form on $\mathbb{Z}$ is $Q(k\alpha) = (1-m)k^2$. $\Delta(S_0) = \{\pm\alpha\}$ and $\Delta(S_m) = (\mathbb{Z}\setminus\{0\})\alpha$ if $m > 0$. $W(S_0) = \{r_\alpha, 1\}$ and $W(S_m) = \{1\}$ if $m > 0$. S_m is of finite, tame or hyperbolic type iff $m = 0$, $m=1$ or $m > 1$ respectively.

b) Denote by P_m the graph with two vertices and m edges connecting these vertices. The associated quadratic form on $\mathbb{Z}^2$ is $Q(k_1\alpha_1+k_2\alpha_2) = k_1^2 - mk_1k_2 + k_2^2$. The set of positive roots is as follows:

$\Delta_+(P_1) = \{\alpha_1, \alpha_2, \alpha_1 + \alpha_2\}$;

$\Delta_+(P_2) = \{k\alpha_1 + (k-1)\alpha_2, (k-1)\alpha_1 + k\alpha_2, k\alpha_1 + k\alpha_2;\ k \geq 1\}$

the set $\{k\alpha_1 + k\alpha_2; k \geq 1\}$ being $\Delta_+^{im}(\Gamma)$;

$m \geq 3$: $\Delta_+(P_m) = \{k_1\alpha_1 + k_2\alpha_2 \mid k_1^2 - m\,k_1k_2 + k_2^2 \leq 1,\ k_1 \geq 0, k_2 \geq 0, k_1+k_2 \geq 0\}$;

more explicitly,

$\Delta_+^{re}(P_m) = \{c_j\alpha_1 + c_{j+1}\alpha_2, c_{j+1}\alpha_1 + c_j\alpha_2,\ j \in \mathbb{Z}_+\}$, where $c_j\ (j \in \mathbb{Z}_+)$ are defined by the recurrent formula:

$$c_{j+2} = mc_{j+1} - c_j,\quad c_0 = 0,\ c_1 = 1.$$

$W(P_1)$ is the dihedral group of order 6 and $W(P_m)$ is the infinite dihedral group if $m \geq 2$. P_m is of finite, tame or hyperbolic type iff $m = 1$, $m = 2$ or $m \geq 3$ respectively.

c) Denote by V_m the graph:

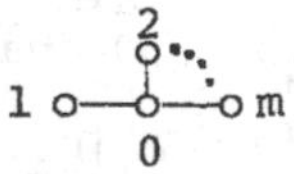

Note that $V_1 = P_1$. The associated quadratic form on $\mathbb{Z}^{m+1}$ is:

$$Q(k_0\alpha_0+\ldots+k_m\alpha_m) = \sum_{i=0}^{m} k_i^2 - k_0\sum_{i=1}^{m} k_i.$$

$\Delta(V_2) = \{\alpha_0, \alpha_1, \alpha_2, \alpha_0 + \alpha_1, \alpha_0 + \alpha_2, \alpha_0 + \alpha_1 + \alpha_2\}$.

$\Delta(V_3) = \{\alpha_0, \alpha_1, \alpha_2, \alpha_3, \alpha_0 + \alpha_1, \alpha_0 + \alpha_2, \alpha_0 + \alpha_3, \alpha_0 + \alpha_1 + \alpha_2, \alpha_0 + \alpha_1 + \alpha_3, \alpha_0 + \alpha_2 + \alpha_3, \alpha_0 + \alpha_1 + \alpha_2 + \alpha_3, 2\alpha_0 + \alpha_1 + \alpha_2 + \alpha_3\}$.

$\Delta^{im}(V_4) = (\mathbb{Z}\setminus\{0\})\delta$, where $\delta = 2\alpha_0 + \alpha_1 + \alpha_2 + \alpha_3 + \alpha_4$;

$\Delta^{re}(V_4) = \{n\delta \pm \alpha_i,\ i = 0,1,\ldots,4;\ n\delta \pm (\alpha_0 + \alpha_{i_1} + \ldots + \alpha_{i_s})$, where $1 \leq i_1 < i_2 < \ldots < i_s$, $1 \leq s \leq 4$; $n \in \mathbb{Z}\}$.

V_m is of finite, tame or wild type iff $m \leq 4$, $m = 4$ or $m \geq 5$ respectively; V_m is hyperbolic iff $m = 5$.

d) Denote by $T_{p,q,r}$ the graph:

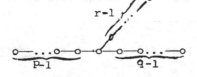

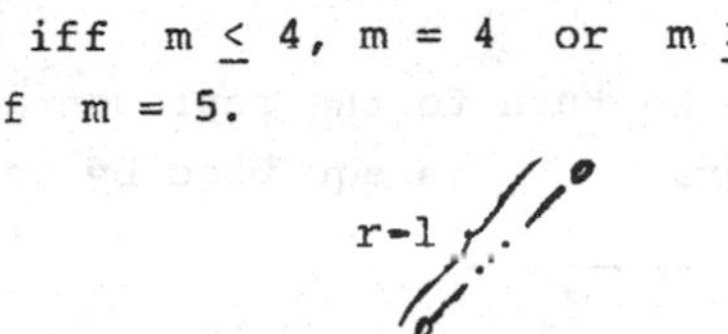

set $c = \frac{1}{p} + \frac{1}{q} + \frac{1}{r}$. Then $T_{p,q,r}$ is of finite, tame or wild type iff $c > 1$, $c = 1$ or $c < 1$ respectively. The only hyperbolic graphs among them are $T_{7,3,2}$, $T_{5,4,2}$ and $T_{4,3,3}$.

§1.4. There are at least two more equivalent difinitions of the set of positive roots $\Delta_+(\Gamma)$:

a) $\Delta_+(\Gamma)$ is a subset of $\mathbb{Z}_+^n\setminus\{0\}$ such that:

(i) $\alpha_1,\ldots,\alpha_n \in \Delta_+(\Gamma)$; $2\alpha_i \notin \Delta_+(\Gamma)$ for $\alpha_i \in \Pi$;

(ii) if $\alpha_i \notin \Pi$, $\alpha \in \Delta_+(\Gamma)$, then $\alpha + \alpha_i \in \Delta_+(\Gamma)$;

(iii) if $\alpha \in \Delta_+(\Gamma)$, then supp α is connected;

(iv) if $\alpha \in \Delta_+(\Gamma)$, $\alpha_i \in \Pi$ and $\alpha \neq \alpha_i$,

then $[\alpha, r_{\alpha_i}(\alpha)] \cap \mathbb{Z}^n \subset \Delta_+(\Gamma)$.

b) Assume that Γ has no edges-loops. Extend the action of the group $W(\Gamma)$ to the lattice $\mathbb{Z}^n \oplus \mathbb{Z}\rho$ by: $r_{\alpha_i}(\rho) = \rho - \alpha_i$. Then one can show that $s(w) := \rho - w(\rho) \in \mathbb{Z}_+^n\setminus\{0\}$ for all $w \in W$. For $w \in W$ set $\varepsilon(w) = \det(w)$; it is clear that $\varepsilon(w) = \pm 1$. Introduce the notation: $x^\alpha = x_1^{k_1}\ldots x_n^{k_n}$, where $\alpha = \sum_i k_i\alpha_i$, and take the product decomposition of the following sum:

$$\sum_{w\in W} (\det w)x^{s(w)} = \prod_{\alpha\in\mathbb{Z}_+^n} (1 - x^\alpha)^{m_\alpha}.$$

Then one can show that $m_\alpha \in \mathbb{Z}_+$, and that

$$\Delta_+(\Gamma) = \{\alpha \in \mathbb{Z}_+^n \mid m_\alpha > 0\}.$$

The positive integer m_α is called the multiplicity of the root α. Note that $m_\alpha = m_{w(\alpha)}$ for $w \in W$.

I do not know how to extend this definition to the case when Γ has edges-loops.

Examples. The multiplicity of a real root is 1. The multiplicity of an imaginary root of a tame graph $\tilde{A}_n$, $\tilde{D}_n$ or $\tilde{E}_n$ is n. This gives the multiplicity of any root α such that $(\alpha,\alpha) = 0$ since any such root is W-equivalent to a unique imaginary root β such that supp β is a tame graph. One knows that if $(\alpha,\alpha) < 0$, then mult $k\alpha$ growth exponentially as $k \to \infty$.

§1.5. Now we turn to the representation theory of quivers. If every edge of a graph Γ is equipped by an arrow, we say that Γ is equipped by an orientation, say Ω; an oriented graph (Γ,Ω) is called a quiver.

Fix a base field $\mathbb{F}$. A representation of a quiver (Γ,Ω) is a collection of finite dimensional vector spaces V_j, $j = 1,\ldots,n$, and linear maps $\phi_{ij}: V_i \longrightarrow V_j$ for every arrow $i \longrightarrow j$ of the quiver (Γ,Ω), everything defined over $\mathbb{F}$. The element $\alpha = \sum_i (\dim V_i)\alpha_i \in \mathbb{Z}_+^n$ is called the <u>dimension</u> of the representation. Morphisms and direct sums of representations of (Γ,Ω) are defined in an obvious way (the dimension of a direct sum is equal to the sum of dimensions). A representation is called <u>indecomposable</u> (resp. <u>absolutely indecomposable</u>) if it is not zero and cannot be decomposed into a direct sum of non-zero representations defined over $\mathbb{F}$ (resp. $\overline{\mathbb{F}}$, the algebraic closure of $\mathbb{F}$).

The main problem of the theory is to classify all representations of a quiver up to isomorphism. One knows that the decomposition of a representation into indecomposable ones is unique. So, for classification purposes it is sufficient to classify indecomposable representations.

Note that there exists a unique up to isomorphism representation of dimension $\alpha_i (i = 1,\ldots,n)$ and it is absolutely indecomposable, namely: $V_i = \mathbb{F}$, $V_j = 0$ for $j \neq i$ and all the maps are zero.

§1.6. Examples.

a) The graph S_m has a unique orientation. The problem of classification of the representations of this quiver is equivalent to the classification of m-tuples of $k \times k$-matrices up to a simultaneous conjugation by a non-degenerate matrix. For $m = 1$ this problem is "tame" and was solved by Weierstrass and Jordan (the so called Jordan normal form). For $m \geq 2$ the problem remains open and provides a typical example of a "wild" problem.

b) Put on P_m the orientation Ω for which all arrows point into the same direction. The corresponding problem is to classify all m-types of linear maps from one vector space into another. For $m = 1$ this is a trivial "finite" problem. For $m = 2$ this is a "tame" problem, which was solved by Kronecker. For $m \geq 3$ the problem becomes "wild".

c) Put on the graph V_m the orientation Ω for which all arrows point to the vertex 0. The corresponding problem is essentially equivalent to the problem of classification of m-tuples of subspaces in a vector space V up to an automorphism of V. For $m \leq 3$ the problem is "finite". For $m = 4$ the problem is "tame" and was solved by Nazarova and Gelfand-Ponomarev, for $m \geq 5$ the problem becomes "wild".

Now I shall give precise definitions. A quiver is called <u>finite</u> if it has only a finite number of indecomposable representations (up to isomorphism). Following Nazarova [10], we call a quiver (Γ,Ω) <u>wild</u> if there is an imbedding of the category of representations of the quiver S_2 into the category of representations of (Γ,Ω); a quiver which is not finite or wild is called <u>tame</u>.

Gabriel [2] proved that the quiver (Γ,Ω) is finite iff Γ is finite (i.e., appears in Table F); this will follow from our general theorems. Nazarova [10] proved that (Γ,Ω) is tame iff Γ is tame (i.e., appears in Table T).

Let me show on examples how to prove that (Γ,Ω) is wild.

For the quiver from c) take a vector space V, put $V_1 = V_2 = V$, and take the 1'st map $V_1 \longrightarrow V_2$ to be an isomorphism. Then the category of representations of S_{m-1} is naturally imbedded in the category in question. So (P_m,Ω) is wild if $m \geq 3$.

For the quiver from c) take $V_0 = V \oplus V$, $V_1 = V_2 = \ldots = V_m = V$. Let $A_1, A_2, \ldots, A_{m-3}: V \longrightarrow V$ be some linear operators. Define the maps $\phi_i: V \longrightarrow V \oplus V$ $(i=1,\ldots,m)$ by:

$$\phi_i(x) = x \oplus A(x) \quad \text{for} \quad i = 1,\ldots,m-3;$$

$$\phi_{m-2}(x) = x \oplus 0, \ \phi_{m-1}(x) = 0 \oplus x, \ \phi_m(x) = x \oplus x.$$

It is easy to see that this is an imbedding of the category of representation of S_{m-3} in the category in question. So the quiver (V_m,Ω) is wild if $m \geq 5$.

§1.7. One of the main technical tools of the representation theory of quivers are the so called reflection functors. Given an orientation Ω of a graph Γ and a vertex k, define a new orientation $\tilde{r}_k(\Omega)$ of Γ by reversing the direction of arrows along all the edges containing the vertex k. A vertex k of a quiver (Γ,Ω) is called a <u>sink</u> (resp. <u>source</u>) if for all edges for which k is a vertex, the arrows point to the vertex k (resp. to the other vertex). Note that if there is a loop at k, then k is neither a sink nor a source.

<u>Proposition</u> [1]. <u>Let</u> (Γ,Ω) <u>be a quiver and</u> k <u>a sink (resp. source)</u>. <u>Then there exists a functor</u> R_k^+ <u>(resp.</u> R_k^-<u>)</u> <u>from the category of representations of the quiver</u> (Γ,Ω) <u>to the category of representation of the quiver</u> $(\Gamma,\tilde{r}_k(\Omega))$ <u>such that</u>:

a) $R_k^{\pm} (U \oplus U') = R_k^{\pm}(U) \oplus R_k^{\pm}(U')$;

b) <u>If</u> U <u>is a representation of dimension</u> α_k, <u>then</u> $R_k^{\pm}(U) = 0$;

c) If U is an indecomposable representation of (Γ,Ω) and $\dim U \neq \alpha_k$, then

$$R_k^- R_k^+(U) \simeq U \quad (\text{resp. } R_k^+ R_k^-(U) \simeq U),$$
$$\text{and } \dim R_k^{\pm}(U) = r_{\alpha_k}(\dim U).$$

Corollary. Under the assumptions of c) of the proposition, $R_k^{\pm}(U)$ is an indecomposable representation of $(\Gamma, \check{r}_k(\Omega))$, and $\operatorname{End} U$ and $\operatorname{End} R_k^{\pm}(U)$ are canonically isomorphic.

We shall explain the construction of the reflection functors R_k^+ and R_k^- in the next section in a more general situation.

§1.8. Now we establish a link between representation theory of quivers and invariant theory.

Fix $\alpha = \sum_i k_i \alpha_i \in \mathbb{Z}_+^n$. Then the set of all, up to isomorphism, representations of dimension α of the quiver (Γ,Ω) is in 1-1 correspondence with the orbits of the group

$$G^\alpha(\mathbb{F}) := GL_{k_1}(\mathbb{F}) \times \ldots \times GL_{k_n}(\mathbb{F})$$

operating in a natural way on the vector space

$$M^\alpha(\Gamma,\Omega) := \bigoplus_{i \to j} \operatorname{Hom}_{\mathbb{F}}(\mathbb{F}^{k_i}, \mathbb{F}^{k_j})$$

(here the summation is taken over all arrows of the quiver (Γ,Ω)).

Note that the subgroup $C = \{(t,\ldots,t), t \in \mathbb{F}^*\}$ operates trivially and that

$$(=) \qquad \dim G^\alpha - \dim M^\alpha(S,\Omega) = (\alpha,\alpha).$$

Furthermore, note that $U \in M^\alpha(\Gamma,\Omega)$ is an indecomposable representation of the quiver (Γ,Ω) iff $\operatorname{End} U$ contains no nontrivial projectors (recall that P is a projector if $P^2 = P$). U is an absolutely indecomposable representation iff $\operatorname{End} U$ contains no nontrivial semisimple elements, i.e., iff the stabilizer $(G^\alpha/C)_U$ of $U \in M^\alpha(S,\Omega)$ is a unipotent group.

Another observation: the group G_U^α is connected since it is the set of invertible elements in the ring End U.

Now I shall explain, what are the reflection functors. Let G be a group and π_1, π_2 some representations of G on vector spaces V_1 and V_2, $\dim V_1 = m \geq k$. Then the group $G \times GL_k$ acts naturally on the space

$$M^+ = \operatorname{Hom}(V_1, \mathbb{F}^k) \oplus V_2.$$

Set $M_0^+ = \{\phi \oplus v \in M^+ \mid \phi \in \operatorname{Hom}(V_1,\mathbb{F}^k),\ v \in V_2,\ \operatorname{rank}\phi = k\}$. Furthermore, the group $GL_{m-k} \times G$ acts naturally on $M^- = \mathrm{Hom}(\mathbb{F}^{m-k}, V_1) \oplus V_2$. We set $M_0^- = \{\phi \oplus v \in M^- \mid \phi \in \mathrm{Hom}(\mathbb{F}^{m-k}, V_1), v \in V_2, \text{ rank } \phi = m-k\}$. We define a map R^+ from the set of orbits on M_0^+ to the set of orbits on M_0^- as follows. If $\phi \oplus v$ lies on an orbit $\sigma \subset M_0^+$, choosing an isomorphism $\mathbb{F}^{m-k} \longrightarrow \mathrm{Ker}\,\phi$, we get a map $r^+(\phi): \mathbb{F}^{m-k} \longrightarrow V_1$; denote by $R^+(\sigma)$ the orbit of $r^+(\phi) \oplus v \in M_0^-$. It is easy to see that R^+ is a well-defined map. Similarly we define the "dual" map R^- from the set of orbits on M_0^- to the set of orbits on M_0^+. One easily checks that R^-R^+ (resp. R^+R^-) is an identity map on the set of orbits in M_0^+ (resp. M_0^-).

Many people have discovered independently from each other this type of construction. For example, Sato and Kimura call it the "castling transform".

In order to get the reflection functor R_j^+ we apply the above construction to the group $G = \prod_{i \neq j} GL_{k_i}$, $k = k_j$, $V_1 = \bigoplus_{s \to j} \mathrm{Hom}_{\mathbb{F}}(\mathbb{F}^{k_s}, \mathbb{F}^{k_j})$, $V_2 = \bigoplus_{\substack{s \to i \\ i \neq j}} \mathrm{Hom}_{\mathbb{F}}(\mathbb{F}^{k_s}, \mathbb{F}^{k_j})$.

§1.9. We need a general remark about actions of a connected algebraic group G. Let G act on an irreducible algebraic variety X over field $\mathbb{F}$. Then by a theorem of Rosenlicht, there exists a dense open subset $X_0 \subset X$, an algebraic variety Z and a surjective morphism $X_0 \longrightarrow Z$, everything defined over $\mathbb{F}$, whose fibers are G-orbits. Z is called a geometric quotient of X_0.

Now, given an action of G on a constructible set X we can decompose X into a union of irreducible subsets and take a (finite) set of G-invariant algebraic subvarieties $Y_1, \ldots, Y_s \subset X$ such that $\dim X \setminus (Y_1 \cup \ldots \cup Y_s) < \dim X$ and each Y_i has a geometric quotient Z_i. Next, we apply the same procedure to $X \setminus \{Y_1 \cup \ldots \cup Y_s\}$, etc. After at most $\dim X$ steps we obtain (absolutely) irreducible varieties Z_1, $Z_2, \ldots$. We set $\mu(G,X) = \max_i \dim Z_i$. It is clear that this number is well defined. We say that the set of orbits of G on X depends on $\mu(G,X)$ parameters.

Denote by $M_{\mathrm{ind}}^{\alpha}(\Gamma,\Omega)$ the set of all absolutely indecomposable representations from $M^{\alpha}(\Gamma,\Omega)$. This is a G^{α}-invariant set, which is constructible and defined over the prime field. Indeed, there exists a finite number of projectors $P_1, \ldots, P_s$ such that $M_{\mathrm{ind}}^{\alpha}(\Gamma,\Omega) = M^{\alpha}(\Gamma,\Omega) \setminus (\bigcup_i G^{\alpha}(M^{\alpha}(\Gamma,\Omega)^{P_i}))$. Applying the above construction we obtain that the set of absolutely indecomposable representations (considered up to isomorphism) is parametrized by a finite union of algebraic varieties $Z_1, \ldots Z_2, \ldots,$ defined over the prime field. We denote for short:

$$\mu_{\alpha}(\Gamma,\Omega) = \mu(G^{\alpha}, M_{\mathrm{ind}}^{\alpha}(\Gamma,\Omega)).$$

§1.10 Now we can state the main theorem.

Theorem. Suppose that the base field $\mathbb{F}$ is algebraically closed. Let (Γ,Ω) be a quiver. Then

a) There exists an indecomposable representation of dimension $\alpha \in \mathbb{Z}_+^n\backslash\{0\}$ iff $\alpha \in \Delta_+(\Gamma)$.

b) There exists a unique indecomposable representation of dimension α iff $\alpha \in \Delta_+^{re}(\Gamma)$.

c) If $\alpha \in \Delta_+^{im}(\Gamma)$, then $\mu_\alpha(\Gamma,\Omega) = 1-(\alpha,\alpha) > 0$.

The proof of the theorem is based on two lemmas. We defer their proof to the next sections.

Lemma 1. Suppose that α lies in the fundamental set M and that, moreover, $(\alpha,\alpha_i) < 0$ for some i. Then

a) The set $M_0^\alpha(\Gamma,\Omega)$ of representations in $M^\alpha(\Gamma,\Omega)$ with a trivial endomorphism ring is a dense open G^α-invariant subset. In particular, $\mu(G^\alpha, M_0^\alpha(\Gamma,\Omega)) = 1-(\alpha,\alpha)$.

b) $\mu(G^\alpha, M_{ind}^\alpha(\Gamma,\Omega)\backslash M_0^\alpha(\Gamma,\Omega)) < 1-(\alpha,\alpha)$.

Lemma 2. The number of indecomposable representation of dimension α (if it is finite) and $\mu_\alpha(\Gamma,\Omega)$ are independent of the orientation Ω.

Proof of the theorem. Note that using the reflection functors, $\mu_{r_i(\alpha)}(\Gamma,\Omega) = \mu_\alpha(\Gamma,\Omega)$ if $\alpha \neq \alpha_i$ and i is a sink or a source of the quiver (Γ,Ω) (the same is true for the number of indecomposable representations). But using Lemma 2, we can always make the vertex i a sink provided that there is no loops at i. Hence the above statement always holds if there is no loops at i.

If $\alpha \in \Delta_+^{im}(\Gamma)$, by the above remarks we can assume that $\alpha \in M$. If $(\alpha,\alpha_i) = 0$ for all i, then $\operatorname{supp}\alpha$ is a tame graph and $\alpha = k\delta$ (see §1.2), and case by case analysis in [10] gives the result. Now the part c) of the theorem follows from Lemma 1.

Similarly, part b) of the theorem follows from the (trivial) fact that there exists a unique up to isomorphism representation whose dimension is equal to a fundamental root.

To prove c) take $\alpha \in \mathbb{Z}_+^n\backslash W(\Pi)$, $\alpha \neq 0$, and suppose that there exists an indecomposable representation of dimension α. Then, as before, there exists an indecomposable representation of dimension $r_\gamma(\alpha)$ for $\gamma \in \Pi$; in particular, $r_\gamma(\alpha) \in \mathbb{Z}_+^n$. Also $\operatorname{supp}\alpha$ is connected. Hence $W(\alpha) \subset \mathbb{Z}_+^n\backslash W(\Pi)$. Taking $\beta \in W(\alpha)$ of minimal height, we have:

$(\beta,\alpha_i) \leq 0$ for all $\alpha_i \in \Pi$ and supp β is connected. Hence $\beta \in M$ and $\alpha \in \Delta_+^{im}(\Gamma)$. □

Remark. One can show (see e.g. [4]) that a generic representation of dimension $k\delta$ of a tame quiver decomposes into a direct sum of k representations of dimension δ.

§1.11. In this section we prove Lemma 1. Let first $\alpha = \Sigma k_i\alpha_i$ be an arbitrary non-zero element from $\mathbb{Z}_+^n$. Let $\alpha = \beta_1+\ldots+\beta_s$, where $\beta_1 \geq \beta_2 \geq \ldots$ (i.e., each coordinate $\geq$) be a decomposition of α into a sum of non-zero elements from $\mathbb{Z}_+^n$; let $\beta_k = \sum_i m_i^{(k)} \alpha_i$. Taking distinct elements $\lambda_1,\lambda_2,\ldots \in \mathbb{F}^*$ defines a conjugacy class of semi-simple elements in G^α consisting of the elements $g = (g_1,\ldots,g_n)$ such that λ_j is an eigenvalue of g_i with multiplicity $m_i^{(j)}$ for all $i = 1,\ldots,n$. Denote by $S_{\beta_1,\ldots,\beta_s} \subset G^\alpha$ the union of all such conjugacy classes. Then an easy computation shows that the dimensions of the centralizer G_g^α of $g \in G^\alpha$ and of the fixed point set $M^\alpha(\Gamma,\Omega)^g$ of g in $M^\alpha(\Gamma,\Omega)$ are independent of the choice of $g \in S_{\beta_1,\ldots,\beta_s}$ and, moreover, we have:

$$(!) \qquad \dim G_g^\alpha - \dim M^\alpha(\Gamma,\Omega)^g = \sum_i(\beta_i,\beta_i).$$

It follows from the theory of sheets in GL_k[6] that $S_{\beta_1,\ldots,\beta_s}$ is a locally closed irreducible subvariety in G^α; denote by $\hat{S}_{\beta_1,\ldots,\beta_s}$ the union of orbits of the same dimension in the Zarisky closure of $S_{\beta_1,\ldots,\beta_s}$. Then, as we saw, $S_{\beta_1,\ldots,\beta_s} \subset \hat{S}_{\beta_1,\ldots,\beta_s}$. Futhermore, it follows from the theory of sheets in GL_k [6] that $\hat{S}_{\beta_1,\ldots,\beta_s}$ contains a unique unipotent conjugacy class u, which corresponds to the conjugate partition of α. A similar (but slightly more delicate computation, which can be found in [4]) shows that the above properties hold for $g = u$. By a deformation argument it follows that these properties hold for arbitrary $g \in G^\alpha$ (this also can be checked by a direct computation, cf §1.13). So, we have proved the following

Lemma. <u>For</u> $g \in \hat{S}_{\beta_1,\ldots,\beta_s}$, <u>dimensions of</u> G_g^α <u>and</u> $M^\alpha(\Gamma,\Omega)^g$ <u>are independent of</u> g <u>and formula</u> (!) <u>holds</u>.

Note that $\hat{S}_{\beta_1,\ldots,\beta_s}$ is a sheet in G^α, i.e., an irreducible component of the union of the orbits of G^α of the same dimension, so that G^α is a disjoint union of the sets $\hat{S}_{\beta_1,\ldots,\beta_s}$ [6]. Note also that the trivial sheet $\hat{S}_\alpha$ coincides with C.

We need one more lemma. Its proof is based on the following identity:

$$(\#)\qquad \sum_{i,j=1}^{n} a_{ij}m_i(k_j-m_j) = \sum_{j=1}^{n} m_j(k_j-m_j)k_j^{-1}\Big(\sum_{i=1}^{n} a_{ij}k_j\Big) + \tfrac{1}{2}\sum_{i,j=1}^{n} a_{ij}\Big(\frac{m_i}{k_i}-\frac{m_j}{k_j}\Big)^2 k_ik_j,$$

provided that $a_{ij} = a_{ji}$ and $k_j \neq 0$ for all $i,j = 1,\dots,n$. This can be checked directly.

Lemma. Let $\alpha \in M$. Then

$$(!!)\qquad \dim G^{\alpha} - \dim G^{\alpha}_{g} \leq \dim M^{\alpha}(\Gamma,\Omega) - \dim M^{\alpha}(\Gamma,\Omega)^{g}.$$

The equality holds only in the following situation:

(N) $(\alpha,\alpha_i) = 0$ for all $i \in \operatorname{supp}\alpha$ or $g \in C$.

Proof. Using formula (=) from §1.8 and formula (!), we have only to show that $(\alpha,\alpha) \leq \sum_i (\beta_i,\beta_i)$ and the equality holds only in the situation (N). This is equivalent to: $\sum_i (\alpha-\beta_i,\beta_i) \leq 0$ and the equality holds only on the situation (N). We can assume that $\operatorname{supp}\alpha = \Gamma$. Applying identity (#) we deduce:

$$(\alpha-\beta_t,\beta_t) = \sum_j m_j^{(t)}(k_j-m_j^{(t)})k_j^{-1}(\alpha,\alpha_i) + \tfrac{1}{2}\sum_{i,j}(\alpha_i,\alpha_j)\Big(\frac{m_i^{(t)}}{k_i}-\frac{m_j^{(t)}}{k_j}\Big)^2 k_ik_j.$$

Since $(\alpha_i,\alpha_j) \leq 0$ for $i \neq j$ and $(\alpha,\alpha_i) \leq 0$, we deduce that both summands of the right-hand side are ≤ 0. This proves the inequality in question. In the case of equality, both summands are zero. Since the second summand is zero and Γ is connected, we deduce that α and β_t are proportional. Since $\alpha \neq \beta_t$, and the first summand is zero, we deduce that $(\alpha,\alpha_i) = 0$ for all i. □

Now we can easily complete the proof of Lemma 1. Indeed, if $g \in G^{\alpha}\setminus C$, then, by inequality (!!) we have:

$$\dim M^{\alpha}(\Gamma,\Omega) > \dim M^{\alpha}(\Gamma,\Omega)^{g} + (\dim G^{\alpha} - \dim G^{\alpha}_{g}).$$

It follows that $\dim M^{\alpha}(\Gamma,\Omega) > \dim(G^{\alpha}(M^{\alpha}(\Gamma,\Omega)^{g}))$ and therefore there exists a dense open set $M(g)$ in $M^{\alpha}(\Gamma,\Omega)$ such that the intersection of the conjugacy class of g with G^{α}_{U} is trivial for any $U \in M(g)$. Since there exists only a finite number of conjugacy classes of projectors in $\bigoplus_i gl_{k_i}(\mathbb{F})$, we deduce that there is a dense open set M'

in $M^{\alpha}(\Gamma,\Omega)$ such that $(G^{\alpha}/C)_U$ is a unipotent group for any $U \in M'$. Since there is only a finite number of unipotent classes in G^{α}_u we deduce that there is a dense open set in $M^{\alpha}(\Gamma,\Omega)$ which consists of representations with a trivial endomorphism ring. This proves Lemma 1a).

To prove b) note that $\mu(M^{\alpha}_{ind}(\Gamma,\Omega)\setminus M^{\alpha}_0(\Gamma,\Omega)) \leq \max_u(\dim M^{\alpha}(\Gamma,\Omega))^u - \dim(G^{\alpha}/C)_u$ where u ranges over a set of representatives of all non-trivial unipotent classes of G^{α}. But the right-hand side is (by (!!)) $< \dim M^{\alpha}(\Gamma,\Omega) - \dim G^{\alpha} + 1$, which is equal to $1-(\alpha,\alpha)$. □

§1.12. Unfortunately, I do not know a direct proof of Lemma 2. The only known proof requires a reduction mod p argument and counting over a finite field. In this section we recall the necessary facts.

Let X be an absolutely irreducible N-dimensional algebraic variety over a finite field $\mathbb{F}_q$ of $q = p^s$ elements (p is a prime number). Then the number of points in X over the field $\mathbb{F}_{q^t}$ is equal to $q^{Nt} + \phi(t)$, where $\phi(t)/q^{Nt} \longrightarrow 0$ as $t \longrightarrow \infty$. This is a simplest fact of the Weil philosophy. In other words knowing the number of points in X over all finite fields $\mathbb{F}_{q^t}$ we can compute the dimension of X.

Let now X be an absolutely irreducible N-dimensional algebraic variety over $\mathbb{Q}$. Then X can be represented as a union of open affine subvarieties, each of which is given by a system of polynomial equations over $\mathbb{Z}$, the transition functions being polynomials over $\mathbb{Z}$. Now we can reduce this modulo a prime p. Then for all but a finite number of primes we get an absolutely irreducible variety $X^{(p)}$ over $\mathbb{F}_p$ of dimension N.

This reduces the proof of Lemma 2 to the case when $\mathbb{F}$ is a field of prime characteristic p.

§1.13. In order to count the number of orbits of $G^{\alpha}(\mathbb{F}_q)$ on $M^{\alpha}(\Gamma,\Omega)(\mathbb{F}_q)$ we employ the Burnside lemma: for the action of a finite group G on a finite set Y the number of orbits is:

$$|Y/G| = \frac{1}{|G|}\sum_{g\in G}|Y^g|.$$

(Here Y^g denote the fixed point set of g on Y and $|Z|$ denotes the cardinality of Z). Denoting by C_g the congugacy class of $g \in G$ and using $|C_g| = |G|/|G_g|$ we can rewrite this formula:

$$|Y/G| = \sum_g |Y^g|/|G_g|,$$

where the summation is taken over a set of representatives of conjugacy classes in G.

Now we need a Jordan canonical form for the elements from $GL_k(\mathbb{F}_q)$ (this information can be found, e.g., in [9], Chapter IV).

Denote by Φ the set of all irreducible polynomials in t over $\mathbb{F}_q$ with leading coefficient 1, excluding the polynomial t. Such a polynomial of degree d has the form

$$P(t) = \prod_{i=0}^{d-1} (t - \alpha^{q^i}), \quad \text{where} \quad \alpha \in \mathbb{F}_q^*, \alpha^{q^d} = \alpha.$$

It follows that the number of polynomials from Φ of degree d is equal to

$$q - 1 \quad \text{if} \quad d = 1 \quad \text{and} \quad d^{-1} \sum_{j|d} \mu(j) q^{d/j} \quad \text{if} \quad q > 1,$$

where μ denotes the classical Möbius function.

Let Par denote the set of all partitions, i.e., non-increasing finite sequences of non-negative integers: $\lambda = \{\lambda_1 \geq \lambda_2 \geq \ldots\}$. We denote by λ' the conjugate partition and by $m_i(\lambda)$ the multiplicity of i in λ; we denote: $|\lambda| = \sum_i \lambda_i$, $\langle\lambda,\mu\rangle = \sum_i \lambda_i \mu_i$.

Conjugacy classes C_ν in $GL_k(\mathbb{F}_q)$ are parametrizes by maps ν: $\Phi \longrightarrow \text{Par}$ such that $\sum_{P \in \Phi} (\deg P) |\nu(P)| = k$ as follows.

To each $f = t^d - \sum_{i=1}^{d} a_i t^{i-1}$ we associate the "companion matrix"

$$J(f) = \begin{bmatrix} 0 & 1 & 0 & \ldots & 0 \\ 0 & 0 & 1 & \ldots & 0 \\ \ldots & \ldots & \ldots & \ldots & \ldots \\ 0 & 0 & 0 & \ldots & 1 \\ a_1 & a_2 & a_3 & \ldots & a_d \end{bmatrix},$$

and for each integer $m \geq 1$ let

$$J_m(f) = \begin{bmatrix} J(f) & 1 & 0 & \ldots & 0 \\ 0 & J(f) & 1 & \ldots & 0 \\ \ldots & \ldots & \ldots & \ldots & \ldots \\ 0 & 0 & 0 & \ldots & J(f) \end{bmatrix}$$

with m diagonal blocks $J(f)$. Then the Jordan canonical form for elements of the conjugacy class C_ν is the diagonal sum of matrices $J_{\nu(f)_i}(f)$ for all $i \geq 1$ and $f \in \Phi$.

The order of the centralizer of each $g \in C_\nu$ is

$$a_\nu(q) = q^{\sum_{P\in\Phi} (\deg P) \langle \nu(P)', \nu(P)' \rangle} \prod_{P\in\Phi} b_{\nu(P)}(q^{-\deg P})$$

where for $\lambda \in \mathrm{Par}$, $b_\lambda(q) = \prod_{i\geq 1} (1-q^{-1})(1-q^{-2})\dots(1-q^{-m_i(\lambda)})$

Finally, if $g \in C_\nu$ and $h \in C_\gamma \subset GL_m$, then (see e.g. [4]):

$$\dim(\mathbb{C}^k \otimes \mathbb{C}^m)^{g\otimes h} = \sum_{P\in\Phi} (\deg P) \quad \langle \nu(P)', \gamma(P)' \rangle.$$

Let now (Γ,Ω) be an oriented graph, and $\alpha = \Sigma k_i\alpha_i \in \mathbb{Z}_+^n$. The conjugacy classes of G^α are parametrized by the maps $\Phi \longrightarrow \mathrm{Par}^n$. An element $\lambda \in \mathrm{Par}^n$ is an n-tuple of partitions $\lambda^{(i)} = \{\lambda_1^{(i)} \geq \lambda_2^{(i)} \geq \dots\}$; set $\lambda_j = (\lambda_j^{(1)},\dots,\lambda_j^{(n)}) \in \mathbb{Z}_+^n$. For $\lambda,\mu \in \mathrm{Par}^n$ we define $(\lambda,\mu) = \sum_j (\lambda_j,\lambda_j)$, where the bilinear form $(\ ,\ )$ on $\mathbb{Z}^n$ is the one associated to Γ. This pairing depends on the graph Γ but is independent of Ω.

Using the Burnside formula we easily deduce the following formula for the number of orbits $d_\alpha(q)$ of $G^\alpha(\mathbb{F}_q)$ on $M^\alpha(\Gamma,\Omega)(\mathbb{F}_q)$:

$$d_\alpha(q) = \sum_\nu \frac{q^{-\sum_{P\in\Phi}(\deg P)\,(\nu',\nu')}\, q}{\prod_{k=1}^{n} \prod_{P\in\Phi} b_{\nu(P)(k)}\ (q^{\deg P})},$$

where ν ranges over all maps $\nu\colon \Phi \longrightarrow \mathrm{Par}^n$ such that $\sum_{P\in\Phi} (\deg P)\,|\nu(P)^{(i)}| = k_i$. This formula (derived jointly with R. Stanley) is quite intractible. However, the following two important corollaries of this formula are clear:

The number $d_\alpha(q)$ of isomorphism classes of representations of the quiver (Γ,Ω) over $\mathbb{F}_q$ is independent of the orientation Ω, and is a polynomial in q with rational coefficients.

(It is immediate that $d_\alpha(q)$ is a rational function in q over $\mathbb{Q}$, but since $d_\alpha(q) \in \mathbb{Z}$ for all $q = p^s$, p prime, $s \in \mathbb{Z}_+$, it follows, that, in fact, $d_\alpha(q)$ is a polynomial.)

We deduce by induction on $\mathrm{ht}\,\alpha$ the following

<u>Lemma.</u> <u>The number of isomorphism classes of indecomposable representations of dimension</u> α <u>of the quiver</u> (Γ,Ω) <u>over the field</u> $\mathbb{F}_q$ <u>is a polynomial in</u> q <u>with rational coefficients, independent of the orientation</u> Ω.

§1.14. It remains to pass from indecomposable representations to absolutely indecomposable ones. For that we need the following general result, the proof of which can be found e.g., in [13] (see also [3]).

Proposition. *Let* G *be a connected algebraic group operating transitively on an algebraic variety* X *over a finite field* $\mathbb{F}_q$. *Suppose that the stablizer* G_x *of* $x \in X$ *is connected. Then the set* $X(\mathbb{F}_q)$ *of points defined over* $\mathbb{F}_q$ *is non-empty and* $G(\mathbb{F}_q)$ *operates transitively on it.*

Since all the stabilizers of the action of G^α on $M^\alpha(\Gamma,\Omega)$ are connected, by the proposition, counting the points over $\mathbb{F}_q$ of the geometric quotients $Z_1, Z_2, \ldots$ is the same as counting the orbits of $G^\alpha(\mathbb{F}_q)$ in $M^\alpha_{ind}(\Gamma,\Omega)(\mathbb{F}_q)$.

In order to count the number of absolutely indecomposable representations over $\mathbb{F}_q$ we need the following lemma, which follows easily from the proposition (see [3],p.90).

Lemma. a) *A representation* U *of* (Γ,Ω), *defined over a finite field* $\mathbb{F}$, *has a unique minimal field of definition* $\mathbb{F}'$. *If* $\sigma \in \mathrm{Gal}(\mathbb{F}':\mathbb{F}_p)$ *and* U *is isomorphic to* U^σ, *then* $\sigma = 1$.

b) *Let* $U \in M^\alpha(\Gamma,\Omega)$ *be an absolutely indecomposable representation of* (Γ,Ω) *with a finite minimal field of definition* $\mathbb{F}'$. *Let* $\mathbb{F}_q \subset \mathbb{F}'$ *and set* $G = \mathrm{Gal}(\mathbb{F}':\mathbb{F}_q)$. *Set* $\tilde{U} = \bigoplus_{\sigma\in G} U^\sigma$. *Then*

(i) $\tilde{U} \in M^{n\alpha}(\Gamma,\Omega)$ *is indecomposable over* $\mathbb{F}_q$ *and* $\mathbb{F}_q$ *is the minimal field of definition for* $\tilde{U}$;

(ii) *two such representations* $\tilde{U}$ *and* $\tilde{V}$ *are isomorphic over* $\mathbb{F}_q$ *iff* U *is isomorphic over* $\mathbb{F}'$ *to a G-conjugate of* V;

(iii) *every indecomposable representation for which* $\mathbb{F}_q$ *is the minimal field of definition can be obtained in the way described above.*

Now we can easily finish the proof of Lemma 2 (and of the theorem). Denote by $m(\Gamma,\alpha;q)$ (resp. $m'(\Gamma,\alpha,q)$) the set of absolutely indecomposable (resp. indecomposable) representations over $\mathbb{F}_q$ of dimension α of the quiver (Γ,Ω). Then we deduce from the lemma that for an indivisible $\alpha \in \mathbb{Z}_+^n$ one has:

(n) $$m'(\Gamma, r\alpha; q) = \sum_{d|r} \frac{1}{d} \sum_{k|d} \mu(k)\, m(\Gamma, \tfrac{r}{d}\alpha; q^{\frac{d}{k}}),$$

where μ is the classical Möbius function. From this one expresses

$m(\Gamma,r\alpha;q)$ via $m'(\Gamma,d\alpha;q^s)$ where $d|r$. Hence $m(\Gamma,r\alpha;q)$ is independent of Ω. □

§1.15. Note that we have also the following

<u>Proposition.</u> $m(\Gamma,\alpha;q) = q^{\mu_\alpha} + a_1 q^{\mu_\alpha - 1} + \ldots + a_{\mu_\alpha}$, <u>where</u> $\mu_\alpha = 1-(\alpha,\alpha)$ <u>and</u> $a_1, a_2, \ldots$ <u>are integers, independent of</u> Ω <u>and</u> q; <u>moreover</u>, $m(\Gamma,w(\alpha);q) = m(\Gamma,\alpha;q)$ <u>for any</u> $w \in W$.

<u>Proof.</u> It follows from the remarks in §1.12 and the main theorem that

$$m(\Gamma,\alpha;q^t) = q^{\mu_\alpha t} + \phi(t), \quad \text{where} \quad \phi(t)/q^{\mu_\alpha t} \longrightarrow 0 \quad \text{as} \quad t \longrightarrow \infty.$$

On the other hand, by §1.13, $m'(\Gamma,\alpha;q)$ and hence $m(\Gamma,\alpha;q)$ is a polynomial in q with rational coefficients. Since $m(\Gamma,\alpha;q) \in \mathbb{Z}$ for all $q = p^s$, it follows that the coefficients are integers. The rest of the statements were proved in the previous sections. □

<u>Conjecture</u> 1. $a_{\mu_\alpha} = \text{mult}\ \alpha$ (provided that Γ has no edges-loops).

<u>Conjecture</u> 2. $a_i \geq 0$ for all i.

I have no idea what is the meaning of the rest of a_i's.

<u>Examples</u>. a) If $\alpha \in \Delta_+^{re}$, then $m(\Gamma,\alpha;q) = 1$.

b) If Γ is a tame quiver with $n+1$ vertices, and $\alpha \in \Delta_+^{im}$, then $m(\Gamma,\alpha;q) = q + n$.

c) Let (Γ,Ω) be the quiver V_k from §1.6 and let $\beta_k = 2\alpha_0 + \alpha_1 + \alpha_2 + \ldots + \alpha_k \in \Delta_+(V_k)$. Then one can show (using Peterson's reccurent formula) that the multiplicity of β_k satisfies the following reccurent relation:

$$(k-1)(\text{mult}\ \beta_k) = k(\text{mult}\ \beta_{k-1}) + 2^{k-2}(k-2); \ \text{mult}\ \beta_3 = 1.$$

From this we deduce: $\text{mult}\ \beta_k = 2^{k-1} - k$.

On the other hand one has (as D. Peterson pointed out):

$$m(V_k,\beta_k) = (q+1)^{k-3} + 3m(V_{k-1},\beta_{k-1}) - 2m(V_{k-2},\beta_{k-2}),$$

which gives:

$$m(V_k,\beta_k) = q^{k-3} + \binom{k}{1}q^{k-4} + \left(\binom{k}{2} + \binom{k}{0}\right)q^{k-5} + \left(\binom{k}{3} + \binom{k}{1}\right)q^{k-6} + \left(\binom{k}{4} + \binom{k}{2} + \binom{k}{0}\right)q^{k-7} + \ldots$$

and $m(V_k,\beta_k;0) = 2^{k-1} - k$.

All the examples agree with the conjectures!

Remark. The constant terms of the polynomials $m(\Gamma,\alpha;q)$ and $m'(\Gamma,\alpha;q)$ are equal. Indeed, by formula (α) we have:

$$m'(\Gamma,r\alpha;0) = \sum_{d|r} \frac{1}{d} \sum_{k|d} \mu(k)m(\Gamma,\frac{r}{d}\alpha;0) = \sum_{d|r} \frac{1}{d} m(\Gamma,\frac{r}{d}\alpha;0)(\sum_{k|d} \mu(k)) =$$

$$= m(\Gamma,r\alpha;0), \quad \text{since} \quad \sum_{k|d} \mu(k) = 0 \quad \text{unless} \quad d = 1.$$

Conjecture 2 naturally suggests one more

Conjecture 3. The set of isomorphism classes of indecomposable representations of a quiver admits a cellular decomposition by locally closed subvarieties isomorphic to affine spaces, a_i's being the number of cells of dimension $\mu_\alpha - i$.

It follows from the proofs that the minimal field of definition of the (unique) representation of (Γ,Ω) of dimension $\alpha \in \Delta_+^{re}$ is $\mathbb{F}_p$ if char $\mathbb{F} = p$. Ironically enough, I do not know how to prove

Conjecture 4. If char $\mathbb{F} = 0$, the representation of (Γ,Ω) of dimension $\alpha \in \Delta_+^{re}$ is defined over $\mathbb{Q}$.

It would be interesting to give an explicit construction of this representation.

More general is the following

Conjecture 5. The main theorem holds over an arbitrary field $\mathbb{F}$.

Note that it is clear that if $\alpha \notin (\mathbb{Z}_+\Delta_+^{re}) \cup \Delta_+^{im}$, then there is no indecomposable representations of dimension α over $\mathbb{F}$. It would follow from Conjecture 4 that this is the case also for $\alpha \notin \Delta_+$.

§1.16. It is easy to see (see [3]) that the theorem, as well as Conjectures 4 and 5 would follow from the following

Conjecture 6. Let G be a linear algebraic group operating on the vector space V, all defined over $\mathbb{F}$, such that char $\mathbb{F} = 0$ or char $\mathbb{F} > \dim V$. Denote by V_0 (resp V_0^*) the sets of points with a unipotent stabilizer in V (resp. V^*). Then the number of orbits of G on V_0 is equal to that of G on V_0^* and $\mu(G,V_0) = \mu(G,V_0^*)$.

Example: $\mathbb{F} = \mathbb{R}$, $G = \{\begin{pmatrix} a & b \\ 0 & 1 \end{pmatrix}$, where $a > 0\}$, $V = \mathbb{R}^2$, action on V

(resp. V*) is the multiplication on a vector-column from the left (resp. vector row from the right). For the action on V the orbits of $\binom{1}{0}$ and $\binom{-1}{0}$ are the two (1-dimensional) orbits with a (1-dimensional) unipotent stabilizer. For the action of G on V* the orbits of (10) and (-10) are the two open orbits with a trivial stabilizer.

The following generalization of Conjecture 6 was suggested by Dixmier.

Conjecture 7. Let $S \subset G$ be a reductive subgroup of G. Denote by V_S (resp. V_S^*) the set of points $x \in V$ (resp. $\in V^*$) such that a Levi factor of G_x is a conjugate of S. Then the numbers of orbits of G in V_S and V_S^* are equal and $\mu(G,V_S) = \mu(G,V_S^*)$.

Remark. It is easy to deduce from Conjecture 6 the following statement: Fix a miximal torus $T \subset G$ and denote by V_T(resp. V_T^*) the set of $x \in V$ (resp. $\in V^*$) such that a maximal torus of G_x is a conjugate of T. Then the numbers of orbits of G in V_T and V_T^* are equal and $\mu(G,V_T) = \mu(G,V_T^*)$.

More general is the following

Conjecture 8. Let N be the unipotent radical of G; G/N acts on the sets of orbits V/N and V*/N. These two actions are equivalent.

§1.17. Examples. a) If (Γ,Ω) is a finite type quiver there is no imaginary roots and we recover Gabrial's theorem: $U \longmapsto \dim U$ gives a 1 - 1 correspondence between the set of isomorphism classes of indecomposable representations of a finite type quiver (Γ,Ω) and the set $\Delta_+(\Gamma)$.

We consider in more detail the finite type quiver V_3, which corresponds to the problem of classification of triples of subspaces U_1,U_2,U_3 in a given vector space U_0 up to an automorphism of U_0. There are 12 roots in $\Delta_+(\Gamma)$. Apart from the roots $\alpha_1,\alpha_2,\alpha_3$ which correspond to $U_0 = 0$ we have 9 nontrivial indecomposable triples. The corresponding dimensions ($\dim U_0$; $\dim U_1$, $\dim U_2$, $\dim U_3$) are

(1;0,0,0), (1;1,0,0), (1;0,1,0), (1;0,0,1), (1;1,1,0), (1;1,0,1), (1;0,1,1), (1;,1,1,1) and (2;1,1,1).

Let $a_0,a_1,\ldots,a_8$ be the number of times these representations appear as indecomposable direct summands in a given representation $(U_0;U_1,U_2,U_3)$. Then we have:

$$\sum_{i=0}^{7} a_i + 2a_8 = \dim U_0, \quad \sum_{i=1}^{8} a_i + 2a_8 = \dim(V_1 + V_2 + V_3),$$

$$a_1 + a_4 + a_5 + a_7 + a_8 = \dim V_1, \quad a_2 + a_4 + a_6 + a_7 + a_8 = \dim V_2,$$

$$a_3 + a_5 + a_6 + a_7 + a_8 = \dim V_3, \quad a_4 + a_7 = \dim V_1 \cap V_2,$$

$$a_5 + a_7 = \dim V_1 \cap V_3, \quad a_6 + a_7 = \dim V_2 \cap V_3, \quad a_7 = \dim V_1 \cap V_2 \cap V_3.$$

It is clear from this system of equation that the nine discrete parameters $\dim U_i (i = 0,\dots,4)$, $\dim U_i \cap U_j$ $(i,j = 1,2,3,\ i = j)$, $\dim U_1 \ U_2 \ U_3$ and $\dim(U_1 + U_2 + U_3)$ determine the triple of subspaces U_1, U_2, U_3 is the vector space U_0 up to isomorphism.

b) The quiver P_2 from §1.6 corresponds to the problem of classification of pairs of linear maps $A,B\colon V_1 \longrightarrow V_2$, the problem solved by Kronecker. We assume that $\mathbb{F}$ is algebraically closed. Then the complete list of indecomposable pairs is in some bases $\{e_i\}$ and $\{f_i\}$ of V_1 and V_2 as follows $(k = 1,2,\dots)$:

$\dim V_1 = k$, $\dim V_2 = k + 1$:

$A(e_i) = f_i; B(e_i) = f_{i+1} (i = 1,\dots,k)$.

$\dim V_1 = k + 1$, $\dim V_2 = k$:

$A(e_i) = f_i (i = 1,\dots,k)$, $A(e_{k+1}) = 0$;

$B(e_i) = f_{i-1} (i = 2,\dots,k + 1)$, $B(e_1) = 0$.

$\dim V_1 = \dim V_2 = k$:

$A(e_i) - f_i (i = 1,\dots,k)$: $B(e_i) = \lambda f_i + f_{i+1} (i = 1,\dots,k - 1)$,

$B(e_k) = \lambda f_k$. Here $\lambda \in \mathbb{F}$ is arbitrary .

$A(e_i) = f_{i+1} (i = 1,\dots,k - 1)$, $A(e_k) = 0$;

$B(e_i) = f_i$.

§1.10. Since the problems of classifcation of all representations of an arbitrary quiver (Γ,Ω) seems to be too difficult, we shall try to understand a simpler question: what is the structure of a generic representation of given dimension α. It is easy to see that there exists a unique decomposition

$$\alpha = \beta_1 + \ldots + \beta_s \ , \quad \text{where} \quad \beta_i \in \mathbb{Z}_+^n \setminus \{0\},$$

such that the set $M_0^\alpha(\Gamma,\Omega) := \{U \in M^\alpha(\Gamma,\Omega) \,|\, U \simeq \bigoplus_{i=1}^{s} U_i,\ \dim U_i = \beta_i$ and all U_i are indecomposable$\}$ is a dense open subset in $M^\alpha(\Gamma,\Omega)$. This is called the canonical decomposition of α. Further on we assume the base field $\mathbb{F}$ to be algebraically closed.

In order to study this decomposition we need the following definition. A representation $U \in M^\alpha(\Gamma,\Omega)$ is called a Schur representation if $\operatorname{End} U = \mathbb{F}$ (or, equivalently, $(G^\alpha/C)_U = 1$). An element $\alpha \in \mathbb{Z}_+^n \setminus \{0\}$ is called a Schur root for the quiver (Γ,Ω) if $M^\alpha(\Gamma,\Omega)$ contains a Schur representation. In this case the set of Schur representations form a dense open subset $M_0^\alpha(\Gamma,\Omega)$ in $M^\alpha(\Gamma,\Omega)$. Note that $\alpha = \alpha$ is the canonical decomposition of a Schur root. Conversely, if there exists a dense open subset in $M^\alpha(\Gamma,\Omega)$ consisting of indecomposable representations (i.e., $\alpha = \alpha$ is the canonical decomposition of α), then α is a Schur root. Indeed, otherwise,

$$\mu_\alpha(\Gamma,\Omega) > \dim M^\alpha(\Gamma,\Omega) - \dim G^\alpha + 1 = 1 - (\alpha,\alpha),$$

a contradiction with the statement c) of the main theorem.

The set of Schur roots is a subset in $\Delta_+(\Gamma)$ (by the main theorem); we denote it by $\Delta_+^{\text{Schur}}(\Gamma,\Omega)$. As will be clear from examples, this set (as well as the canonical decomposition) depends on the orientation Ω of the quiver.

Remark. One can show that even a stronger result holds [4]: If a representation $U \in M^\alpha(\Gamma,\Omega)$ is stably indecomposable, i.e., all representations from a neighbourhood of U are indecomposable, then U is a Schur representation (the converse is obvious). The quiver S_1 with relation $A^2 = 0$ shows that this property fails for quiver with relations. It might be interesting to study the rings R which have the property that every its stably indecomposable representation has a trivial endomorphism ring.

Example. Consider in the 3-dimensional space V_0 a quadruple of subspaces V_1,V_2,V_3,V_4 of dimensions 2,2,1,1 respectively. This quadruple is indecomposable iff $V_1 = V_2$, and $V_3 + V_4$ is a 2-dimensional subspace different from V_1 and V_2 and $\dim V_1 \cap V_2 \cap (V_3 + V_4) = 1$ (all such quadruples are equivalent). However, the generic quadruple is decomposable and the canonical decomposition is as follows:

$$\alpha := 3\alpha_0 + 2\alpha_1 + 2\alpha_2 + \alpha_3 + \alpha_4 = (2\alpha_0 + \alpha_1 + \alpha_2 + \alpha_3 + \alpha_4) + (\alpha_0 + \alpha_1 + \alpha_2)$$

So, α is a (real) root but not a Schur root.

§1.19. Given a quiver (Γ,Ω), let r_{ij} denote the number of arrows with the initial vertex i and the final vertex j. We define the (in general non symmetric) bilinear form R (R in honour of Ringel) on $\mathbb{Z}^n$ by [11]:

$$R(\alpha_i,\alpha_j) = \delta_{ij} - r_{ij}.$$

Note that $(\alpha,\beta) = \frac{1}{2}(R(\alpha,\beta) + R(\beta,\alpha))$ is the associated symmetric bilinear form. The following proposition is crucial.

<u>Proposition [11]</u>. <u>Let</u> U <u>and</u> V <u>be representations of a quiver</u> (Γ,Ω) <u>of dimensions</u> α <u>and</u> β <u>respectively</u>. <u>Then</u>

$$\dim \operatorname{Hom}(U,V) - \dim \operatorname{Ext}(U,V) = R(\alpha,\beta).$$

Using formula (=) from §1.8 we deduce the following well-known formula for an arbitrary representation $U \in M^\alpha(\Gamma,\Omega)$:

$$(\equiv) \qquad \dim M^\alpha(\Gamma,\Omega) - \dim G^\alpha(U) = \dim \operatorname{Ext}(U,U).$$

We need another formula, which also can be derived from the proposition by a straight forward computation.

<u>Lemma</u>. <u>Let</u> $U_j \in M_0^{\beta_j}(\Gamma,\Omega)$ $(j = 1,\ldots,s)$ <u>and</u> $\alpha = \sum_{j=1}^{s} \beta_j$. <u>Let</u> $S \in G^\alpha$ <u>be a semisimple element, such that</u> $\alpha = \sum_j \beta_j$ <u>is the corresponding partition of</u> α. <u>Then</u>

$$\dim M^\alpha(\Gamma,\Omega) - \dim G^\alpha \, (M^\alpha(\Gamma,\Omega)^S) = \sum_{i\neq j} \dim \operatorname{Ext}(U_i,U_j)$$

<u>Proof</u>. It is clear that the left-hand side of the formula is equal to:

$\dim M^\alpha(\Gamma,\Omega) - \dim G^\alpha(U) - \dim M^\alpha(\Gamma,\Omega)^S + \dim G^\alpha_S = \dim M^\alpha(\Gamma,\Omega) - \dim G^\alpha +$ $+ \dim G^\alpha_U + \dim G^\alpha_S - \dim M^\alpha(\Gamma,\Omega)^S = -(\alpha,\alpha) + \sum_j (\beta_j,\beta_j) + \dim G^\alpha_U$ by formula (=) from §1.8 and (!) from §1.11. Since $\dim G^\alpha_U = \dim \mathrm{Hom}(U,U)$ the lemma follows from the proposition. □

Corollary. *Let* $U_i \in M_0^{\beta_i}(\Gamma,\Omega)$ *and* $U = \oplus_i U_i$, $\alpha = \sum_i \beta_i$. *Then* $U \in M_0^\alpha(\Gamma,\Omega)$ *iff* $\mathrm{Ext}(U_i,U_j) = 0$ *for all* $i \neq j$. *In particular, if all* β_i *are Schur roots, then* $\alpha = \sum_i \beta_i$ *is the canonical decomposition of* α.

§1.20. We call an element $\alpha \in \mathbb{Z}^n_+\setminus\{0\}$ *indecomposable* if α cannot be decomposed into a sum $\alpha = \beta + \gamma$, where $\beta,\gamma \in \mathbb{Z}^n_+\setminus\{0\}$ and $R(\beta,\gamma) \geq 0$, $R(\gamma,\beta) \geq 0$. One deduces immediately from the remarks in §1.17 and the corollary from §1.18 the following facts:

Proposition. a) *If* α *is an indecomposable element, then* α *is a Schur root.*

b) *Let* $\alpha \in \mathbb{Z}^n_+\setminus\{0\}$ *and* $\alpha = \beta_1+\ldots+\beta_s$ *be the canonical decomposition of* α. *Then all* β_i *are Schur roots and* $R(\beta_i,\beta_j) \geq 0$ *for all* $i \neq j$.

Conjecture 9. If α is a Schur root, then α is indecomposable.

Conjecture 10. Provided that (Γ,Ω) has no oriented cycles, each $\alpha \in \mathbb{Z}^n\setminus\{0\}$ admits a unique decomposition $\alpha = \sum_j \beta_j$ such that β_j are indecomposable and $R(\beta_i,\beta_j) \geq 0$ for $i \neq j$. (See [4] for a version of this conjecture without assumptions on (Γ,Ω)).

If the conjectures 9 and 10 were true, we obtain that the decomposition of α given by conjecture 10 coincides with its canonical decomposition.

In [4] conjectures 9 and 10 are checked for finite and tame quivers, and for rank 2 quivers.

Example. If (Γ,Ω) is finite, then conjecture 9 holds since then any root is indecomposable. Indeed, if $\alpha = \beta + \gamma$, where $R(\beta,\gamma) \geq 0$ and $R(\gamma,\beta) \geq 0$, then $1 = (\alpha,\alpha) = (\beta,\beta) + (\gamma,\gamma) + R(\beta,\gamma) + R(\gamma,\beta) \geq 2$. Similarly, we show that if (Γ,Ω) is a tame quiver and α is a root such that its defect $R(\delta,\alpha) \neq 0$ then α is indecomposable.

Remark. If $(\alpha,\alpha_i) \leq 0$ for all i and $(\alpha,\alpha_i) < 0$ for some i, then α is indecomposable by the identity (#), and hence is a Schur root. This gives another proof of Lemma 1a).

§1.21. The following simple facts proved in [4] show that many questions about the action of G^α on $M^\alpha(\Gamma,\Omega)$ can be answered in terms of the canonical decomposition.

Proposition. Let $\alpha \in \mathbb{Z}_+^n\setminus\{0\}$ and let $\alpha = \beta_1+\ldots+\beta_k$ be the canonical decomposition of α.

a) $\operatorname{tr\,deg} \mathbb{F}(M^\alpha(\Gamma,\Omega))^{G^\alpha} = \sum_{i=1}^{k} (1-(\beta_i,\beta_i))$.

b) $\operatorname{tr\,deg} \mathbb{F}(M^\alpha(\Gamma,\Omega)^{(G^\alpha,G^\alpha)}) = \sum_{i=1}^{k} (1-(\beta_i,\beta_i)) + |\operatorname{supp}\alpha| - s - r$,

where s and r are the number of distinct real roots and the dimension of the $\mathbb{Q}$-span of all imaginary roots in the canonical decomposition of α, respectively.

c) G^α has a dense orbit in $M^\alpha(\Gamma,\Omega)$ iff all β_i are real, the principal stabilizer being reductive iff $R(\beta_i,\beta_j) = 0$ whenever $\beta_i \neq \beta_j$.

d) If G^α has a dense orbit $\mathcal{O}$ in $M^\alpha(\Gamma,\Omega)$, then we have for the categorical quotient:

$$M^\alpha(\Gamma,\Omega)/(G^\alpha,G^\alpha) \simeq \mathbb{F}^{|\operatorname{supp}\alpha| - s},$$

where s is the same as in b) (and also is the number of distinct indecomposable summands of a representation from $\mathcal{O}$).

e) The generic (G^α,G^α)-orbit in $M^\alpha(\Gamma,\Omega)$ is closed iff $R(\beta_i,\beta_j) \neq 0$ whenever $\beta_i \neq \beta_j$.

Remarks. a) If (Γ,Ω) is a finite type quiver, then G^α always has a dense orbit in $M^\alpha(\Gamma,\Omega)$ (since it has a finite number of orbits or by part c) of the proposition), and hence formula from d) always holds. This has been found by Happel.

b) If (Γ,Ω) is a tame quiver then G^α has a dense orbit in $M^\alpha(\Gamma,\Omega)$ iff the defect $R(\delta,\alpha) \neq 0$ (by part c) of the proposition).

Chapter II. The slice method.

The slice method is based on Luna's slice theorem [7] and was for the first time applied in [5] for the classification of irreducible representations of connected simple linear groups for which the ring of invariants is a polynomial ring. In this chapter I discuss some examples of applications of this method, mainly to invariant theory of binary forms.

§2.1. Let G be a linear reductive group operating on a finite-dimensional vector space V, both defined over $\mathbb{C}$. For $p \in V$ let G_p denote the stabilizer of p and T_p the tangent space to the orbit $G(p)$ of p. Then T_p is G_p-invariant and we can consider the action of G_p on the vector space $S_p := V/T_p$. If the orbit $G(p)$ is closed, the action of G_p on the space S_p is called a slice representation. Note that G_p is a reductive group (since G/G_p is an affine variety and by Matsushima criterion G/H is affine iff H is a reductive subgroup); therefore, we can identify S_p with a G_p-invariant complementary to T_p subspace in V.

The slice method is based on the following principle:

Given a representation of a reductive group, every its slice representation is "better" than the representation itself.

§2.2. In order to make this principle more precise we have to introduce the so called categorical quotient. Let $\mathbb{C}[V]$ denote the ring of polynomials on V and $R = \mathbb{C}[V]^G$ the subring of G-invariant polynomials. Then it follows from the complete reducibility of the action of G on $\mathbb{C}[V]$ that there exists a linear map $\mathbb{C}[V] \longrightarrow \mathbb{C}[V]^G$, denoted by $f \longmapsto f^\natural$ with the following properties:

(i) if $U \subset \mathbb{C}[V]$ is G-invariant, then $U^\natural \subset U$;

(ii) if $f \in \mathbb{C}[V]^G$, $g \in \mathbb{C}[V]$, then $(fg)^\natural = fg^\natural$.

One immediately deduces the classical fact that the algebra $\mathbb{C}[V]^G$ is finitely generated. Indeed, let $I \subset \mathbb{C}[V]$ be the ideal generated by all homogeneous invariant polynomials of positive degree. By Hilbert's basis theorem, it is generated by a finite number of invariant polynomial, say $P_1,\dots,P_N$. We prove by induction on the degree of a homogeneous polynomial $P \in \mathbb{C}[V]^G$ that P lies in the subalgebra generated by $P_1,\dots,P_N$. We have

$$P = \sum_{i=1}^{N} Q_i P_i, \quad \text{where} \quad \deg Q_i < \deg P.$$

Applying to both sides the operator $\natural$ we get:

$$P = \sum_i Q_i^{\natural} P_i, \quad \text{where} \quad \deg Q_i^{\natural} = \deg Q_i < \deg P,$$

and applying the inductive assumption to $Q_i^{\natural}$ completes the proof.

Denote by V/G the affine variety for which $\mathbb{C}[V]^G$ is the coordinate ring. Then the inclusion $\mathbb{C}[V]^G \hookrightarrow \mathbb{C}[V]$ induces a map $\pi\colon V \longrightarrow V/G$ called the quotient map. The pair $\{\pi, V/G\}$ is called the categorical quotient because it satisfies the following characteristic properties:

(i) the fibers of π a G-invariant;

(ii) if $\pi'\colon V \longrightarrow M$ is a morphism, such that M is in affine variety and the fibers of π' are G-invariant, then there exists a unique map $\psi\colon V/G \longrightarrow M$ such that $\pi' = \psi \circ \pi$.

Note that V/G is a (weighted) cone (i.e., it has a closed imbedding $V/G \hookrightarrow \mathbb{C}^m$, which is invariant under transformations $(c_1,\ldots,c_m) \longrightarrow (t^{s_1}c_1,\ldots,t^{s_m}c_m)$, $t \in \mathbb{C}$, $s_i > 0$).

Note that $\mathbb{C}[V]^G$ is a polynomial ring $\Leftrightarrow$ the vertex of V/G is a regular point $\Leftrightarrow$ V/G is smooth.

§2.3. Here we prove the following classical fact: V/G parametizes the closed orbits, i.e., for each $x \in V/G$, the fiber $\pi^{-1}(x)$ contains a unique closed orbit.

This follows from the following two facts:

(i) if $M_1, M_2 \subset V$ are two closed disjoint G-invariant subvarieties, then there exists an invariant polynomial P which is identically 0 on M_1 and identically 1 on M_2;

(ii) the map π is surjective.

Then the closed orbit in a fiber is an orbit of minimal dimension in the fiber.

For (i), let P_i's and Q_j's be generators of defining ideals I_1 and I_2 for M_1 and M_2. Then by Hilbert's Nullstelensatz, $\{P_i, Q_j\}_{i,j}$ generate $\mathbb{C}[V]$ is an ideal, hence $\sum_i g_i P_i + \sum_j g_j' Q_j = 1$ for some g_i, g_j'. Denoting the first summand by f_1 and second by f_2, we have:

$$f_1 + f_2 = 1, \quad \text{where} \quad f_1 \in I_1, \quad f_2 \in I_2.$$

Applying $\natural$, we get:

$$f_1^{\natural} + f_2^{\natural} = 1, \quad \text{where} \quad f_i^{\natural} \in I_i.$$

Set $P = f_1^{\natural}$; clearly, $P|_{M_1} = 0$ since $P \in I_1$, so $P|_{M_2} = 1$.

If (ii) fails then there exists a maximal ideal $I \subsetneqq \mathbb{C}[V]^G$ which generates $\mathbb{C}[V]$ as an ideal; then we have: $1 = \sum_i f_i P_i$, where $f_i \in \mathbb{C}[V]$, and $P_i \in I$. Applying $\natural$, we get:

$$1 = \sum_i f_i^{\natural} P_i.$$

Hence $I = \mathbb{C}[V]^G$, a contradiction. □

§2.4. Let $p \in V$ be such that the orbit $G(p)$ is closed. We have: $V = T_p \oplus S_p$ (G_p-invariant decomposition). We have by restriction: $\pi: S_p \longrightarrow V/G$, and by the universality property, this can be pushed down to the morphism:

$$\pi_p: \; S_p/G_p \longrightarrow V/G.$$

It is clear that $\dim S_p/G_p = \dim V/G$. From Luna's slice theorem we deduce the following lemma [5].

<u>Lemma.</u> <u>The morphism</u> π_p <u>induces an isomorphism of completions of local rings of</u> $\pi(p) \in V/G$ <u>and the vertex of the cone</u> S_p/G_p

Note that this lemma says, in particular, that π_p is an analytic isomorphism on some neighbourhood of the vertex of S_p/G_p (in complex topology), i.e., this vertex has the same singularity as $\pi(p) \in V/G$. Moreover, it follows from §2.3 that we get all the singularities of V/G in this way.

§2.5. Here are some precise special cases of our general principle. Let $p \in V$ be such that the orbit $G(p)$ is closed. Set $R = \mathbb{C}[V]^G$ and $R_1 = \mathbb{C}[S_p]^{G_p}$.

<u>Proposition.</u> a) <u>If</u> R <u>is a polynomial ring (resp. complete intersection), then</u> R_1 <u>is a polynomial ring (resp. complete intersection) too</u>.

b) <u>Let</u> m <u>(resp.</u> m_p<u>)</u> <u>denote the minimal number of homogeneous generators of</u> R <u>(resp.</u> R_1<u>)</u>. <u>Then</u> $m \geq m_p$.

c) <u>Set</u> $A = \mathbb{C}[z_1,\dots,z_m]$, $A_1 = \mathbb{C}[z_1,\dots,z_{m_p}]$ <u>and let</u>

$$\dots \longrightarrow A^{r_2} \longrightarrow A^{r_1} \longrightarrow R \longrightarrow 0 \quad \text{and} \quad \dots \longrightarrow A^{r_{2p}} \longrightarrow A^{r_{1p}} \longrightarrow R_1 \longrightarrow 0$$

<u>be the minimal free resolutions for</u> R <u>and</u> R_1. <u>Then</u>

$$r_1 \geq r_{1p},\ r_2 \geq r_{2p},\ \dots .$$

<u>Proof</u>. follows from §2.4 and the following remarks. From the point of view of the properties we are interested in

(i) the behaviour of a local ring is the same as the one of its completion;

(ii) the behaviour of the local ring of the vertex of a cone is the same as the one of the coordinate ring of the cone;

(iii) the local ring of the vertex is the "worst" among all the local rings of a cone.

All these statements about local rings are quite simple and can be found e.g. in [14]. □

§2.6. One often can find slice representations for which G_p is a finite group. Then one can apply the Shepard-Todd-Chevalley theorem to check that R is not a polynomial ring. In order to check that R is not a complete intersection one can apply the following result [16].

<u>Proposition</u>. <u>Let</u> G <u>be a finite linear group operating on a vector space</u> V <u>of dimension</u> n. <u>Suppose that</u> $\mathbb{C}[V]^G$ <u>has</u> m <u>generators</u> and that the ideal of relations (i.e., the kernel of the surjection $\mathbb{C}[z_1,\dots,z_m] \longrightarrow \mathbb{C}[V]^G$) <u>is generated by</u> $m - n + s$ <u>elements (note that</u> $m \geq n$ <u>and</u> $s \geq 0$). <u>Then</u> G <u>is generated by those</u> σ <u>such that</u> $\operatorname{rank}(\sigma - I) \leq s + 2$. <u>In particular, if</u> V/G <u>is a complete intersection, then</u> $G = \langle \sigma \in G \mid \operatorname{rank}(\sigma - I) \leq 2 \rangle$.

<u>Proof</u>. Let F_g denote the fixed point set of $g \in G$. Denote by Z the union of all $F_g \subset V$ such that $\operatorname{codim}_V F_g \geq s + 3$. Then G acts on $X := V \setminus Z$ and $X/G = (V/G)\setminus(Z/G)$. Note that V/G is simply connected since, being a cone, it is contractible to the vertex. Furthermore, X/G is simply connected by the following fact, proved by Goresky and Macpherson. Let M be a closed affine subvariety in $\mathbb{C}^m$ of dimension n and suppose that the ideal of M is generated by $m - n + s$ elements. Then if M is simply connected, $M\setminus Y$ is

simply connected for any sibvariety $Y \subset M$ of codimension $\geq s + 3$.

So, G acts on a connected variety X such taht X/G is simply connected. But then $G = \langle G_x | x \in X \rangle$. Indeed, let G_1 denote the right-hand side. Then G/G_1 acts on X/G_1 such that $g \neq e$ has no fixed points. Since $X/G = (X/G_1)/(G/G_1)$, we deduce that $G/G_1 = e$. This completes the proof. □

Note that the same proof gives the "only if" part of the Shepard-Todd-Chevalley theorem.

<u>Remark</u>. If $G \subset SL_2(\mathbb{C})$, then $\mathbb{C}^2/G$ is always a complete intersection (F. Klein). But for $G = \{\begin{pmatrix} \varepsilon & 0 \\ 0 & \varepsilon \end{pmatrix}\}$, where ε is a cube root of 1, $\mathbb{C}^2/G$ is not a complete intersection. Indeed, $u_1 = x^3$, $u_2 = x^2y$, $u_3 = xy^2$, $u_4 = y^3$ is a minimal system of generating invariants, and a minimal system of relations is: $u_1u_4 = u_2u_3$, $u_1u_3 = u_2^2$, $u_2u_4 = u_3^2$. This is the simpliest counter-example to the converse statement of the proposition.

§2.7. In order to apply the slice method one should be able to check that an orbit $G(x)$ is closed. For this one can use the Hilbert-Mumford Richardson criterion, which I will not discuss here, or the following

<u>Proposition</u> [8]. <u>Let G be a reductive group operating on a vector space V, $p \in V$ and $H \subset G_p$ be a reductive subgroup. The normalizer N of H in G acts on the fixed point set L of H in V. The orbit $G(p)$ in V is closed iff the orbit $N(p)$ in L is closed.</u>

Note also that $G(p)$ is closed iff $G^0(p)$ is closed, where G^0 is the connected component of the unity of G.

§2.8. Now we turn to an example of the action of the group $G = SL_2(\mathbb{C})$ on the space V of binary forms fo degree d by

$$\begin{pmatrix} \alpha & \beta \\ \gamma & \delta \end{pmatrix} P(x,y) = P(\alpha x + \beta y, \gamma x + \delta y).$$

Fix the following basis of V_d:

$$v_0 = x^d,\ v_2 = x^{d-1}y, \ldots, v_{d-1} = xy^{d-1},\ v_d = y^d.$$

We consider separately the cases d odd and even.

a) d odd and ≥ 3. Set $p = x^{d-1}y + xy^{d-1}$ and $\varepsilon = \exp \frac{2\pi i}{d-2}$. Then $G_p = \{A_k = \begin{pmatrix} \varepsilon^k & 0 \\ 0 & \varepsilon^{-k} \end{pmatrix}, k - 1,\ldots,d-2\}$ is a cyclic group of order $d - 2$. The fixed point set of G_p is $L = \mathbb{C}v_1 + \mathbb{C}v_{d-1}$; the connected component of the unity of the normalizer of G_p is $N^0 = \{\begin{pmatrix} t & 0 \\ 0 & t^{-1} \end{pmatrix}, t \in \mathbb{C}^*\}$. The orbit $N^0(p)$ is clearly closed. Hence by § 2.7, the orbit

G(p) is closed.

The tangent space T_p to $G.p$ is

$$\left[\mathbb{C}x \frac{\partial}{\partial y} + \mathbb{C}y \frac{\partial}{\partial x} + \mathbb{C}(x \frac{\partial}{\partial x} - y \frac{\partial}{\partial y})\right](p) ,$$

hence the eigenvalues of A on T_p are : $\varepsilon^2, \varepsilon^{-2}$ and 1.

On the other hand, we have:

$$A_1(v_j) = \varepsilon^{(d-2j)} v_j .$$

Hence the eigenvalues of A_1 on $S_p = V/T_p$ are:

$$1, \varepsilon, \varepsilon^2, \varepsilon^3, \ldots, \varepsilon^{(d-3)} .$$

So, according to our principle, the representation of a cyclic group $H = \langle A_1 \rangle$ of order $d - 2$ on $\mathbb{C}^{d-2}$ acting by $A_1(e_j) = \varepsilon^{-j} e_j$ in some basis $e_1, \ldots, e_{d-2}$, is "better" than the action of $SL_2(\mathbb{C})$ on the space of binary forms of odd degree $d > 1$.

Let $f_1, \ldots, f_{d-2}$ be the basis dual to $e_1, \ldots, e_{d-2}$. Then the monomials $f_1^{k_1} \cdots f_{d-2}^{k_{d-2}}$ such that

$$(*) \quad \sum_j jk_j \equiv 0 \bmod (d-2)$$

form a basis of the space of invariant polynomials for this action of H.

An integral solution of (*) is called positive if all $k_i \geq 0$ and not all of them = 0; a positive solution is called indecomposable if it is not a sum of two positive solutions. It is clear that the minimal number of generating invariant polynomials for the action of H on $\mathbb{C}^{d-2}$ is equal to the number of indecomposable positive solutions of (*). I do not know how to compute this number. However, (as observed by R. Stanley) it is clear that if a solution $(k_1, \ldots, k_{d-2})$ is positive and the left-hand side of (*) is equal to $d - 2$, then this is an indecomposable solution. Also it is clear that $(d - 2)(\delta_{1i}, \ldots, \delta_{d-2,i})$ is an indecomposable solution provided that i and $d-2$ are relatively prime and $i \neq 1$. This gives us the following estimate: (number of positive indecomposable solutions of (*)) $\geq p(d-2) + \phi(d-2) - 1$, where $p(k)$ is the classical partition function and $\phi(k)$ is the number of $1 \leq j \leq k$ relatively prime to k.

For the discussion of the number of relations we need the following defintiion. Let $R = \mathbb{C}[z_1, \ldots, z_m]/I$ be a finitely generated ring, where m is the minimal number of generators; let $n (\leq m)$ be the dimension of R. We say that R <u>requires at least</u> s <u>extra equations</u> if the minimal system of generators of the ideal I contains at least

$m - n + s$ elements.

It is clear by §2.6 that for the action of the cyclic group H the ring of invariants requires at least $d-5$ extra relations. Again, applying the principle, this gives an estimate for the number of relations between invariants of binary forms.

The obtained results are sumarized in this following

Proposition. Let R be the ring of invariant polynomials for the action of $SL_2(\mathbb{C})$ on the space of binary forms of degree $d > 1$, d odd. Then the minimal number of generators of R is $\geq p(d-2)+\phi(d-2)-1$ and R requires at least $d - 5$ extra relations.

A complete information about degrees of polynomials in a minimal system of homogeneous generators of R (it is easy to see that these are well-defined numbers) and the generating relations are known (for odd d) only for $d \leq 5$. Namely, for $d = 3$, R is generated by a homogeneous polynomial of degree 4; for $d = 5$, R is generated by homogeneous polynomials of degrees 4,8,12 and 18 and there is exactly one generating relation.

Note that for $d = 3$ and 5 our low bounds are exact. However, for $d \geq 7$ the low bounds given by the slice method are far from being exact. For instance taking $p = x^7 + y^7$ for $d = 7$ gives the best low bound, which is 17, for the minimal number of generating invariants; it is known, however, that this number lies between 28 and 33 [15].

b) d even and ≥ 4. We take $p = x^d + y^d$ and let $\varepsilon = \exp \frac{2\pi i}{d}$. Then $G_p = \langle \begin{pmatrix} \varepsilon & 0 \\ 0 & \varepsilon^{-1} \end{pmatrix}, \begin{pmatrix} 0 & 1 \\ -1 & 0 \end{pmatrix} \rangle$. The same argument as in a) shows that the orbit $G(p)$ is closed, and our principle gives similar low bounds. In particular, we get that R requires at least $\frac{3}{4}d - 5$ (resp. $\frac{1}{4}(3s + 2) - 5$) extra relations if $4|d$ (resp. $4|d + 2$).

One has a complete information about R (for even d) only for $d \leq 8$. Namely for $d = 2$, R is generated by one polynomial of degree 2; for $d = 4$, R is freely generated by polynomials for degree 2 and 3; for $d = 6$, R is generated by polynomials of degree 2,4,6, 10 and 15 and there is exactly one generating relation; for $d = 8$ R is generated by polynomials of degree 2,3,4,...,10 and requires two extra relations.

One can see that for $d = 4$, 6 and 8 our low bounds are exact.

§2.9. It follows from the results of §2.8 that for the action of $SL_2(\mathbb{C})$ on the space V_d of binary forms of degree d, the ring R of invariant polynomials is a complete intersection iff $d \leq 6$ (and is a polynomial ring iff $d \leq 4$).

Similarly, one can apply §2.7 to the classification of reductive linear groups for which the ring of invariants is a complete intersection. As an example, let us prove the following

Proposition. For the action of $SL_n(\mathbb{C})$ $(n > 1)$ on the space $S^d(\mathbb{C}^n)$ the ring of invariants R is a complete intersection iff either $d \leq 2$, or $n = 2$ and $d \leq 6$, or $n = d = 3$, or $n = 4$, $d = 3$. Moreover if R is a complete intersection but is not a polynomial ring, then $(n,d) = (2,5)$ or $(2,6)$ or $(4,3)$ and R is the coordinate ring of a hypersurface.

Proof. The case $(4,3)$ was worked out by Salmon about a hundred years ago [12]. He showed that R is generated by invariants of degree 8, 16,24,32,40 and 100 with one generating relation. It is well known that in the case $(3,3)$, R is a polynomial ring generated by invarinats of degree 4 and 6. The case $d \leq 2$ is obvious.

In order to show that in the remaining cases R is not a complete intersection take $p = \sum_{i=1}^{n} z_i^d \in S^d(\mathbb{C}^n)$. Then as in §2.8 we show that the orbit of p is closed. Using §§2.5 and 2.6 we deduce that R is not a complete intersection in all cases in question except $(3,4)$. In the last case one should take $p = z_1^3 z_2 + z_2^3 z_3 + z_3^3 z_1$. □

§2.10. Let $G = SL_n(\mathbb{C})$ and V be the direct sum of $m \geq n$ copies of the natural representation of SL_n on $\mathbb{C}^n$. Let $p = (v_1,\ldots,v_m) \in V$, $p \neq 0$. Then the orbit $G(p)$ is closed iff $\operatorname{rank}(v_1 \ldots v_m) = n$; in this case $G_p = \{e\}$. So all non-trivial slice representations are nice. However, if $m > n + 1$, the point $\pi(0)$ is (the only) singular point of V/G.

In other words, for these representations the slice method does not simplify the problem. In fact the slice principle works best of all for irreducible representations.

References.

1. Bernstein I.N., Gelfand I.M., Ponomarev V.A., Coxeter functors and Gabriel's theorem, Russian Math. Surverys 28, 17-32(1973).

2. Gabriel P., Unzerdegbare Darstellunger I., Man. Math. 6, 71-103 (1972).

3. Kac V.G., Infinite root systems, representations of graphs and invariant theory, Invent. Math. 56, 57-92(1980).

4. Kac V.G., Infinite root systems, representations of graphs and invariant theory II, Journal of Algebra 77, 141-162 (1982).

5. Kac V.G., Popov V.L., Vinberg E.B., Sur les groupes lineares algebrique dont l'algebre des invariants est libre, C.R. Acad. Sci., Paris 283, 875-878(1976).

6. Kraft H. Parametrisieruing der Konjugationklassen in $s\ell_n$, Math. Ann. 234, 209-220(1978).

7. Luna D., Slices étales, Bull. Soc. Math. France, Memoire 33, 81-105(1973).

8. Luna D., Adhérences d'orbites et invariants, Invent. Math. 29, 231-238(1975).

9. Macdonald I.G., Symmetric functions and Hall polynomials, Clarendon press, Oxford, (1979).

10. Nazarova L.A., Representations of quivers of infinite type, Math. U.S.S.R.-Izvestija Ser math. 7, 752-791(1973).

11. Ringel C.M., Representaions of K-species and bimodules, J. of Algebra, 41, 269-302(1976).

12. Salmon G., A treatise of the analytic geometry of three dimensions, v.2, New York, Chelsea, 1958-1965.

13. Serre J.-P., Cohomologie Galoisienne, Lecture Notes in Math. 5 (1965).

14. Serre J.-P., Algèbre Locale, Lecture Notes in Math 11(1975).

15. Sylvester J.J., Tables of the generating functions and ground forms of the binary duodecimic, with some general remarks..., Amer. J. Math. 4, 41-61(1881).

16. Kac V.G., Watanabe K., Finite linear groups whose ring of invariants is a complete intersection, Bull. Amer. Math. Soc. 6 ,221-223(1982).

Victor G. Kac
M.I.T.
Cambridge, MA 02139

INVARIANTS OF Z/pZ IN CHARACTERISTIC p.

Gert Almkvist

1. Representations of $G = \mathbb{Z}/p\mathbb{Z}$.

Let G denote the cyclic group with p elements ($p > 2$ a prime), written multiplicatively. Let further k be a field of characteristic p. Let V be a finite dimensional vector space over k. A representation of G is a group homomorphism

$$\rho : G \to \mathrm{Aut}V.$$

Then V can be considered as a $k[G]$-module via the action

$$g \cdot x = \rho(g)(x)$$

for $g \in G$ and $x \in V$.

The group ring $k[G] \simeq k[X]/(X^p-1)$ is artinian and self-injective (i.e. it is injective considered as a module over itself).

A module V is indecomposable if $V = V_1 \oplus V_2 \Rightarrow V_1 = 0$ or $V_2 = 0$. Then it is not hard to show the following:

1. Every module is a direct sum of indecomposable modules.

2. There exist only finitely many (non-isomorphic) indecomposables $V_1, V_2, \ldots, V_p$ where $\dim_k V_n = n$. Observe that $V_1 = k$ with the trivial G-action and $V_p = k[G]$ is the free module of rank 1.

3. $V_n \simeq k[X]/(X-1)^n$ (let the generator of G act as multiplication by X).

Definition 1.1: The representation ring R_G is the free abelian group with basis $V_1, V_2, \ldots, V_p$ and multiplication defined by $V \otimes_k W$, (G acts by $g\cdot(v \otimes w) = (gv) \otimes (gw)$).

Then $V_1 = 1$ is the identity and we have the multiplication table:

$$V_2 V_n = V_{n+1} + V_{n-1} \quad \text{if } 1 < n < p$$

$$V_2 V_p = 2V_p \ .$$

Hence R_G is generated by V_2 over Z.

More precisely:

<u>Theorem 1.2</u>: The map $V_2 \to X$ induces a ring isomorphism

$$R_G \simeq Z[X] \Big/ (X-2)\, U_{p-1}(X/2)$$

where $U_{p-1}(\cos\varphi) = \frac{\sin p\varphi}{\sin\varphi}$ is the second Chebyshev polynomial of degree $p-1$.

<u>Definition 1.3</u>: Let V be a module. Denote by $V^G = \{x \in V;\ gx = x \text{ for all } g \in G\}$ the <u>submodule of G-invariant elements</u>.

<u>Proposition 1.4</u>: $V_n^G = V_1$.

---

2. <u>Invariants of $G = Z/pZ$</u>.

<u>Definition 2.1</u>: A homogeneous polynomial $f \in k[x_0, x_1, \ldots, x_n]$ is <u>G-invariant</u> if $f(x_0, x_1 + x_0, x_2 + 2x_1 + x_0, \ldots, x_n + \binom{n}{1}x_{n-1} + \binom{n}{2}x_{n-2} + \ldots + x_0) = f(x_0, x_1, \ldots, x_n)$

(i.e. if V has basis $x_0, x_1, \ldots, x_n$ and G has generator σ then σ acts as

$$\sigma = \begin{pmatrix} 1 & 0 & 0 & 0 & \ldots & 0 \\ 1 & 1 & 0 & 0 & \ldots & 0 \\ 1 & 2 & 1 & 0 & \ldots & 0 \\ - & - & - & & & \\ 1 & \binom{n}{1} & \binom{n}{2} & - & - - & 1 \end{pmatrix}$$

If we make the substitution $X \to 1 + X$ it agrees with the action in section 1.

Example 2.2: Let $p = 5$, $n = 3$ and the degree $r = 3$. Then the polynomials

$$x_0^{\,3}$$
$$x_0^{\,2}x_2 - x_0x_1^2$$
$$x_0^2x_3 + 2x_0x_1x_2 + 2x_1^3$$
$$(2x_1^2-2x_0x_2)x_3 + x_1x_2^2 - x_0^2x_1$$

form a basis for the space of homogeneous G-invariants of degree 3, (observe that the first two polynomials agree with the classical $SL(2,\mathbb{C})$-invariants. This is no accident (see [2])).

Main problem of invariant theory: Find the generators and the relations for the graded ring of G-invariants $k[x_0,x_1,\ldots,x_n]^G$. This is very difficult in general. The only non-trivial easy case is the following.

Example 2.3: Let $n = 2$. Then the ring of invariants is the hypersurface

$$k[x_0,x_1,x_2]^G = k[u_0,u_1,u_2,u_3]\Big/\{u_2^2-u_0^pu_3-u_1(u_1^{\frac{p-1}{2}}-u_0^{\,p-1})^2\}$$

where

$$u_0 = x_0$$
$$u_1 = x_1^2 - x_2x_0$$
$$u_2 = x_1^p - x_0^{p-1}x_1$$
$$u_3 = \prod_{j=0}^{p-1}(x_2+2jx_1+j^2x_0)$$

Example 2.4: Let $p = 5$ and $n = 3$ (as in Example 2.2). Then $R = k[x_0,x_1,x_2,x_3]^G$ has twelve generators and at least 16 relations (see [2] and [8]).

If $R = \bigoplus_{r \geq 0} R_r$ is a graded k-algebra where all $\dim_k R_r < \infty$ let

$$F(R,t) = \sum_{r \geq 0} \dim_k R_r t^r$$

denote the <u>Hilbert series of R</u>.

A <u>more reasonable problem</u> is to find the Hilbert series of the ring of invariants.

<u>Example 2.4 (cont.)</u>

$$F(R,t) = \frac{1 - 2t - 2t^2}{(1-t)^3(1-t^5)} = 1 + t + 2t^2 + 4t^3 + \ldots$$

Let V_{n+1} have basis $x_0, x_1, \ldots, x_n$ and let

$$S^r V_{n+1} = \begin{cases} \text{all homogeneous polynomials of} \\ \text{degree } r \text{ in } x_0, x_1, \ldots, x_n \end{cases}$$

be the symmetric product. Then $S^r V_{n+1}$ becomes a G-module via the action

$$g(x_0^{i_0} x_1^{i_n} \ldots x_n^{i_n}) = (gx_0)^{i_0} (gx_1)^{i_1} \ldots (gx_n)^{i_n} .$$

We have

$$k[x_0, x_1, \ldots, x_n]^G = \bigoplus_{r \geq 0} (S^r V_{n+1})^G .$$

If $S^r V_{n+1} = \sum_{j=1}^{p} c_j V_j$ then $\dim_k (S^r V_{n+1}) = \sum_{j=1}^{p} c_j$

because $V_n^G = V_1 = k$.

<u>Theorem 2.5</u>: If n is <u>even</u> then

$$F(k[x_0, \ldots, x_n]^G, t) = \frac{1}{p} \sum_{\gamma \in \mu_p} \prod_{j=0}^{n} \frac{1}{1-\gamma^{n-2j} t}$$

where μ_p = the group of p-th roots of unity.

<u>Remark 2.6</u>: If n is <u>odd</u> there is a more complicated formula (the proof is worse too, see [1] and [3]).

Remark 2.7: The formula in 2.5 looks like Molien's Theorem. Indeed if

$$G' = \left\{ \begin{pmatrix} \gamma^n & & & 0 \\ & \gamma^{n-2} & & \\ & & \ddots & \\ 0 & & & \gamma^{-n} \end{pmatrix} \quad ; \gamma \in \mu_p \right\}$$

acts on $\mathbb{C}[x_0,\ldots,x_n]$ then $\mathbb{C}[x_0,\ldots,x_n]^{G'}$ has the same Hilbert series. But the rings are far from isomorphic.

Reciprocity Theorem 2.8: If n is even then

$$F(k[x_0,\ldots,x_n]^G,\ 1/t) = (-t)^{n+1} F(k[x_0,\ldots,x_n]^G,t)$$

(if n is odd there is no such result).

Amusing(?) Remark 2.9: Put $\gamma = e^{2\pi i\nu/p}$, $\nu = 0,1,2,\ldots,p-1$ and let $p \to \infty$ in 2.5,

$$\frac{1}{p} \sum_{\gamma\in\mu_p} \prod_{j=0} \frac{1}{1-\gamma^{n-2j}t} = \sum_{\nu=0}^{p-1} \frac{1}{\prod_{j=0}^{n} (1-te^{2\pi i\nu(n-2j)/p})} \cdot \frac{1}{p} \to$$

$$\to \frac{1}{2\pi}\int_{-\pi}^{\pi} \frac{1+\cos\varphi}{\prod_{j=0}^{n}(1-te^{i(n-2j)\varphi})}\,d\varphi = \sum_{r\geq 0} b_r t^r = F(C_n,t)$$

(the numerator $1+\cos\varphi$ instead of 1 makes the formula valid also for odd n). Here b_r = number of linearly independent (classical $SL(2,\mathbb{C})$-) covariants of a binary form of degree n, with leading term of degree r (see [2] for an explanation).

Theorem 2.10: If n is even then

$$F(k[x_0,\ldots,x_n]^G,t) =$$

$$-\sum_{0\leq j<n/2} (-1)^j \phi_{n-2j} \left\{ \frac{1+t^p}{1-t^p} \cdot \frac{t^{j(j+1)}}{(1-t)^2\ldots(1-t^{2(n-j)})(1-t^2)\ldots(1-t^{2j})} \right\}.$$

Here $\phi_s : \mathbb{Q}(t) \to \mathbb{Q}(t^s)$ is defined by

$$(\phi_s(h)(t^s) = \frac{1}{s} \sum_{\gamma \in \mu_s} h(\gamma t)$$

(for a more complicated formula when n is odd see [4]).

Example 2.11: Let $n = 4$. Then

$$F(k[x_0,x_1,x_2,x_3,x_4]^G,t) = \frac{(1+t^3)(1+t^p) + 2t^{\frac{p+1}{2}}(1+t)^2}{(1-t)(1-t^2)^2(1-t^3)(1-t^p)}$$

(formulas are also computed for $n = 5$ and 6 for certain classes of primes in [4]).

Ellingsrud and Skjelbred [9] have shown that depth $k[x_0,\ldots x_n]^G = 3$. Since dim $k[x_0,\ldots x_n]^G = n+1$ we have that $k[x_0,\ldots,x_n]^G$ is never Cohen-Macaulay if $n > 2$ (but it is a UFD; see also [7]).

3. Computations in R_G.

The proofs are done in the representation R_G (or rather in the extension $R_G[\mu]$ where $V_2 = \mu + \mu^{-1}$). Also computations are much simplified if one disregards the free modules (i.e. consider the ring $\tilde{R}_G = R_G/(V_p)$).

Definition 3.1: (a) $\lambda_t(V) = \sum_{i\geq 0} \Lambda^i V\, t^i$ in $R_G[t]$

(b) $\sigma_t(V) = \sum_{i\geq 0} S^i V t^i$ in $RG[[t]]$.

Reciprocity Theorem 3.2:

$$\sigma_{1/t}(V_{n+1}) = \begin{cases} (-t)^{n+1}\sigma_t(V_{n+1}) & \text{if } n \text{ is } \underline{\text{even}} \\ (-t)^{n+1}(V_p - V_{p-1})\sigma_t(V_{n+1}) & \text{if } n \text{ is } \underline{\text{odd}}. \end{cases}$$

This is the main result from which all other reciprocity and symmetry theorems follow. [3]

Theorem 3.3 (Symmetry).

(a) $S^r V_{n+1} \simeq S^n V_{r+1}$ if $r,n < p$ ("Hermite's reciprocity law")

(b) $S^r V_{n+1} \simeq \Lambda^r V_{n+r}$ if $n+r < p$

(c) $S^{p-n-r-1} V_{n+1} - S^r V_{n+1} = \frac{1}{p}\left\{\binom{p-r-1}{n} - \binom{r+n}{n}\right\} V_p$ if n is even

$S^{p-n-r-1} V_{n+1} + S^r V_{n+1} = \frac{1}{p}\left\{\binom{p-r-1}{n} + \binom{r+n}{n}\right\} V_p$ if n is odd.

Theorem 3.4

(a) $\lambda_{-t}(V_{n+1})\sigma_t(V_{n+1}) = 1$ if n is even

(b) $\lambda_{-t}(V_{n+1})\sigma_t(V_{n+1}) = \dfrac{1-(V_p - V_{p-1})t^p}{1 - t^p}$ if n is odd.

4. Applications to Combinatorics.

Let $A(m,n,r)$ = number of partitions of m into at most r parts all of size $\leq n$.

Put $V(m,n,r) = A(m,n,r) - A(m-1,n,r)$

Let $A(m,n,r) = 0$ if either $n < 0$, $m > nr$ or $m \notin Z$.

If $S^r V_{n+1} = \sum_{j=1}^{p} c_j V_j$ where $r,n, < p$

then

$$c_j = \sum_{\nu} V\left(\frac{nr+1-j}{2} + \nu p, n, r\right).$$

Hence all results about $S^r V_{n+1}$ (like symmetries) can be formulated in the language of partitions without any reference to group representations.

Let $\begin{bmatrix} n+r \\ r \end{bmatrix}(t) = \frac{(1-t^{n+r}) \ldots (1-t^{n+1})}{(1-t) \quad \ldots (1-t^{r})}$ be the Gaussian polynomial.

Definition 4.1: A polynomial $a_0 + a_1 t + \ldots + a_n t^n$ in $Z[t]$ is

(a) symmetric if $a_j = a_{n-j}$

(b) unimodal if it is symmetric and $0 < a_0 \leq a_1 \leq \ldots \leq a_{[n/2]}$.

Theorem 4.2: Let

$$\begin{bmatrix} n + r \\ r \end{bmatrix}(t) = q(t)(1+t+t^2+ \ldots + t^{p-1}) + t^s f(t)$$

where q and f are symmetric with $f(0) \neq 0$.

Then $f(t)$ is unimodal (see [6]) .

Acknowledgements: My thanks go to R. Fossum who got me interested in invariant theory. Many results are also due to him (see [1]). Furthermore I want to thank L. Avramov and R.P. Stanley for stimulating discussions.

References.

1. G. Almkvist-R. Fossum: Decompositions ... "Sém d'alg., Paul Dubreil 1976-77", Springer Lecture Notes No 641, 1-111.

2. G. Almkvist: Invariants, mostly old ones, Pac. J. Math. 86(1980), 1-13.

3. G. Almkvist: Representations of Z/pZ,..., J.Alg. 68(1981),1-27.

4. G. Almkvist: Some formulas in invariant theory, to appear in J.Alg.

5. G. Almkvist: Rings of invariants, Preprint Lund 1981:2.

6. G. Almkvist: Representations of SL(2,ℂ) and unimodal polynomials, Preprint Lund 1982.

7. M.J. Bertin: Anneaux d'invariants d'anneaux de polynômes en caractéristiques p , CR 264(1967), 653-656.

8. L.E. Dickson: On invariants and the theory of numbers, Dover 1966.

9. G. Ellingsrud-T. Skjelbred: Profondeur d'anneaux d'invariants en caractéristique p , Oslo 1978, No. 1.

University of Lund
Box 725
S-220 07 LUND
Sweden

SYMMETRY AND FLAG MANIFOLDS

by

Alain Lascoux & Marcel-Paul Schützenberger

In spite of its links with the symmetric group, the study of flag varieties has not yet fully used the customary technics (permutoèdre, Ehresman's order, Lehmer's code) of the theory of symmetric functions.

To the so-called Schubert cycles are associated polynomials, which are no other than Schur functions in the case of Grassmann varieties, and which can be studied through the help of symmetrizing operators, acting both on the cohomology ring, and the Grothendieck ring as special cases. Conversely, the study of representations of the symmetric group benefits from the geometrical intuition coming from the action of the symmetric group on the flag variety.

As an example, we indicate how to effectively compute the projective degrees of Schubert cycles. A note submitted to the Acadēmie des Sciences apply these methods to the calculation of the Chern classes of the flag variety - as for its harmonic functions, the theory of which requires some properties of the plactic monoïd, they will be the subject of a separate article.

Half of the authors warmly thanks Mittag Leffler Institute & the University of Stockholm for their hospitality, the C.I.M.E. Foundation for providing the opportunity of displaying the symmetrizing operators, as well as A.Björner, D.Laksov and F.Gherardelli for their interest in this work.

<u>Caution</u> : the operators are placed on the right.

§ 1 Symmetrizing operators.

It is always delicate to distinguish between a permutation and its inverse, or between right and left multiplication for the symmetric group, if one does not take a set of "values" and a set of "places". To avoid misunderstandings, we shall consider permutations of $n+1$ elements as operators on the ring of polynomials $\mathbb{Z}[a,b,\ldots]$, $\{a,b,\ldots\}$ being a totally ordered alphabet of cardinal $n+1$.

Starting from the special element $a^E = a^n\, b^{n-1}\, c^{n-2} \ldots$, one uses the transpositions $\sigma_{ab}, \sigma_{bc} \ldots$ of consecutive letters to generate all monomials (written in the lexicographic order) whose multidegrees are a reordering of $\{0,1,\ldots,n\}$.

This process gives us a ranked poset, as shown in the following figure, the permutations being considered as paths (directed downwards) in the graph of the poset (the graph is called the "permutoëdre"). The permutoëdre gives us all the reduced decompositions of the elements of the symmetric group, the length of a permutation being the length of any corresponding path; ω denotes the permutation of greatest length.

For example, the symmetric group on three letters gives

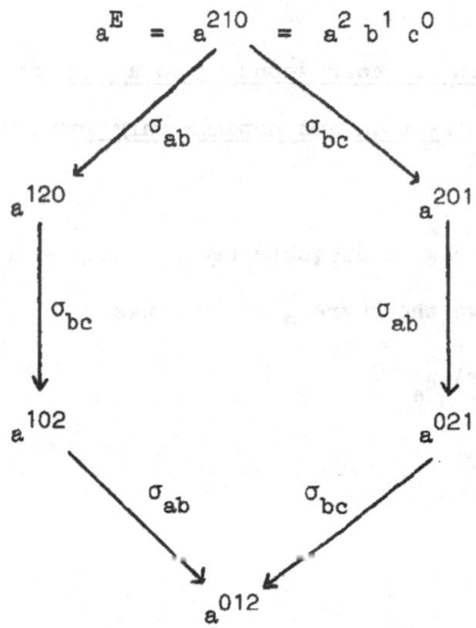

FIGURE 1

Moore's relations say that two paths having the same end points can be obtained by a sequence of elementary transformations

1.1 $\sigma_{ab} \cdot \sigma_{bc} \cdot \sigma_{ab} = \sigma_{bc}\, \sigma_{ab}\, \sigma_{bc}$

a,b,c being any triple of consecutive letters

1.2 $\sigma_{ab} \cdot \sigma_{de} = \sigma_{de}\, \sigma_{ab}$

if $\{a,b\} \cap \{d,e\} = \emptyset$

The conventions are such that if $\sigma . \sigma' . \sigma'' \ldots$ is a path from a^E to $a^I = a^{i_0} b^{i_1} \ldots$, then with $\omega w = i_0+1,\ i_1+1,\ \ldots,\ i_n+1$, one has $a^E\, \omega w = a^I$, $a^I\, w^{-1} = a^0 b^1 c^2 \ldots$; less trivial operators on $\mathbb{Z}[a,b,\ldots]$ appear when applying Jacobi's procedure to generate symmetric polynomials through alternating polynomials.

Define ∂_{ab} to be the operator

1.3 $\partial_{ab} : f(a,b,\ldots) \rightsquigarrow [f(a,b,c\ldots) - f(b,a,c\ldots)]/(b-a)$

and similarly for all pairs of consecutive letters, where f is an arbitrary polynomial (or rational) function.

We can interpret any path of the permutoëdre as a product of operators ∂_{ab}. Checking the relation similar to 1.1 (1.2 is trivial in this case), one gets

1.4 Lemma. *The product of operators corresponding to a path from* ω *to* w *is independant of the choice of the path and depends only upon* w . *It is denoted* $\partial_{\omega w}$.

The operators ∂_w are not always adequate because they systematically decrease the degrees. To preserve the degree, one defines

1.5 $\pi_{ab} : f \rightsquigarrow (af)\, \partial_{ab}$

$\pi_{bc} : f \rightsquigarrow (bf)\, \partial_{bc}$

. . . .

and one checks that these new operators still verify relations 1.1 and 1.2, so that a product of operators corresponding to a path depends only upon the end points : $\pi_{\omega w}$ is given by a path from ω to w , π_w by a path from w^{-1} to 123 ... .

Having at hand three operators verifying the same relations 1.1 and 1.2, one cannot resist in putting them in a single family.

Let p,q,r be fixed integers.

Define

1.6 $$D_{ab}(p,q,r) : f \rightsquigarrow (f)(p\partial_{ab} + q\pi_{ab} + r\sigma_{ab})$$

and similarly for all pairs of consecutive letters.

It is a simple, but not totally trivial verification, that conditions 1.1 and 1.2 are still fulfilled, so that one can write $D_w(p,q,r)$.

To accelerate computations one may remark that symmetric functions are scalars with respect to the D_w :

1.7 $$f,g \in \mathbb{Z}[a,b,..] \ , \ f\sigma_{ab} = f \Rightarrow (fg)D_{ab} = gD_{ab}f \ .$$

In fact $D_\omega(p,q,o)$ is a symmetrizer in the whole alphabet, i.e. $\forall f \in \mathbb{Z}[a,b,...] \ , \ \forall w, [f\, D_\omega(p,q,o)]w = f\, D_\omega(p,q,o)$.

One can show that, up to a change of variables, the operators $D_w(p,q,o)$ are the most general symmetrization operators verifying certain natural conditions, and thus we cannot find a family with more parameters containing them.

More precisely concerning $D_\omega(p,q,o)$, one has:

1.8 $$f\, D_\omega(p,q,o) = \sum (f\, \Delta_{pq})w$$

sum on all permutations, Δ_{pq} being the generalization of Vandermonde's determinant:

$$\Delta_{pq} = \Pi_{x<y} \ (p + qx)/(x - y) \ .$$

1.9 Remark. $f\,\pi_\omega = (fa^E)\partial_\omega = \sum (-1)^{\ell(w)} (fa^E)w \,/\, \sum (-1)^{\ell(w)} a^E w$;

$a^I \pi_\omega$ is the classical Schur function of index $i_n, i_{n-1}, \ldots, i_0$ (cf Macdonald).

Thus the operators $D_\omega(1,0,0) = \partial_\omega$ and $D_\omega(0,1,0)$ are essentially the same, and formula 1.8 becomes in this case Jacobi's expression of Schur functions, Weyl's character formula for the linear group and Bott's theorem for the cohomology of line bundles on flag manifolds.

We did not use the square D^2_{ab} of an operator; in fact, one has

1.10 $$D^2_{ab} = q\, D_{ab} + r(q+r)$$

so that one is really working with a representation of the Hecke algebra of the symmetric group.

§ 2 Schubert polynomials.

As the action of ∂_ω, or π_ω as well, transforms a monomial into a Schur function, the operators D_w will give generalizations of Schur functions.

Following Demazure, and independently, Bernstein-Gelfand and Gelfand, we shall define, for every permutation w, the polynomial X_w by

2.1 $$X_w = a^E \,\partial_{\omega w} ,$$

a^E being the monomial $a^n\, b^{n-1}\, c^{n-2} \ldots$. Thus, the X_w are obtained just by pushing down $X_\omega = a^E$ along the edges of the permutoèdre.

Figure 2 gives the result in the case of four letters.

One could generate the X_w through the π_w ; this more complicated process leads to a combinatorial representation of the X_w .

FIGURE 2

POLYNÔMES DE SCHUBERT

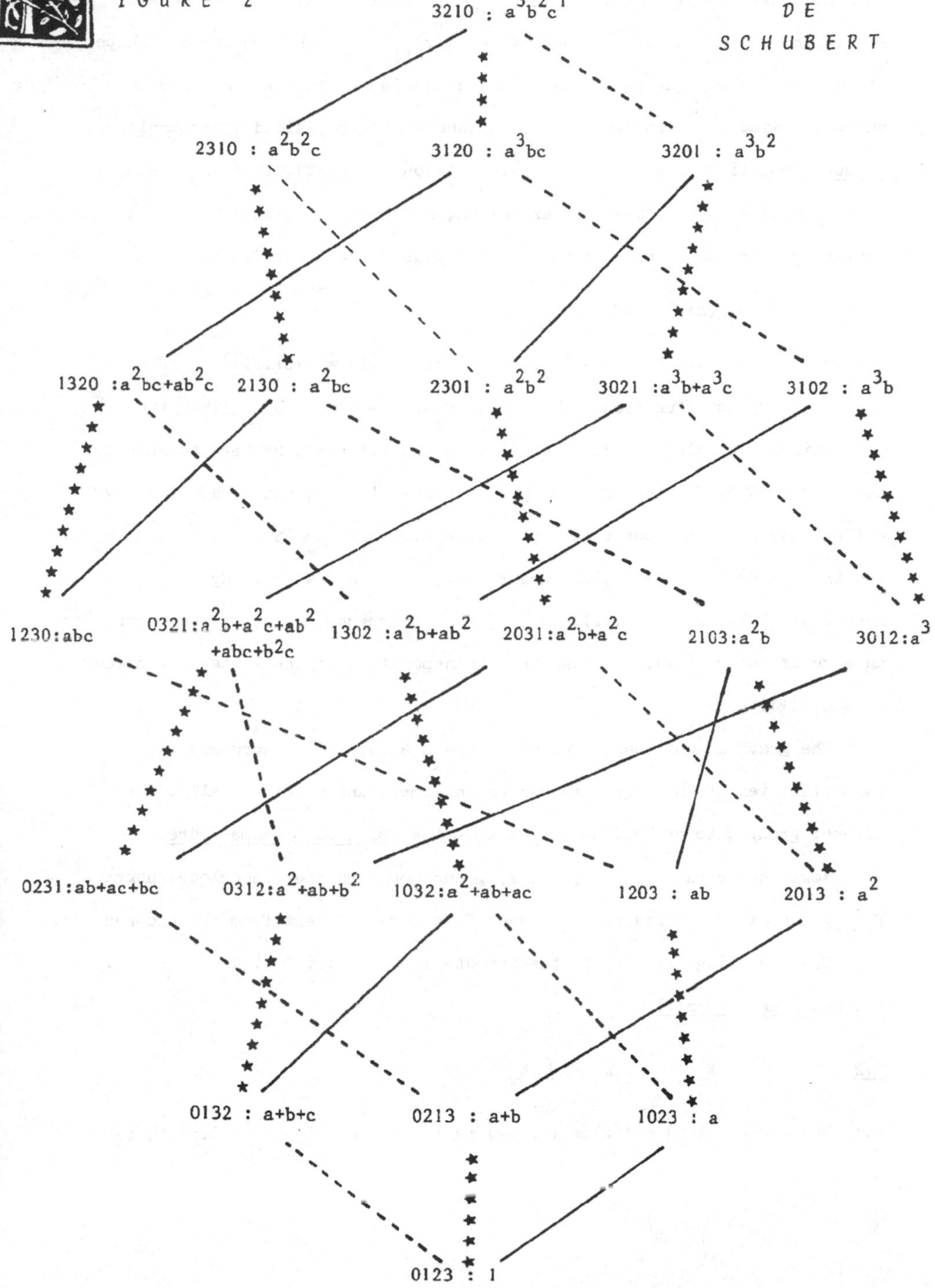

One notices on the example that X_w is a polynomial of degree $\ell(w)$ with positive coefficients, and that X_w is symmetrical in the i-th and i+1-th letters of A if and only if $w_i < w_{i+1}$; in other words, the shape of w (i.e. the sequence of its up's and down's) gives the symmetries of the polynomial X_w . For those permutations which are called grassmannian permutations (i.e. the ones which have only one descent), then X_w is a Schur function in the first letters of A, e.g. X_{2413} , which has to be symmetrical in a,b and also in c,d is indeed the Schur function

$$S_{12}(a+b) = a^2b + ab^2$$

(we identify an alphabet $A = \{a,b,\ldots\}$ and $S_1(A) = a+b+\ldots$) .

As for Schur functions, the first problem will be to multiply two polynomials. The simplest case is due to Monk, but first we need to enlarge the permutoëdre. For each pair of permutations (v,w) , such that v and w differ only by a transposition: $v = \ldots v_i \ldots v_j \ldots \curvearrowright w = \ldots v_j \ldots v_i \ldots$ and that $\ell(w) = \ell(v) + 1$, one draws between v and w j-i edges, of respective "colors" (i+1, i), ..., (j, j-1) (remember that on the permutoëdre, an edge of color (i+1, i) meant the transposition of the letters at place i and i+1).

The graph so obtained, when one forgets about the colours and the multiplicities of the edges, is due to Ehresman, and more generally, for Coxeter groups, to Bruhat. Let us call it the coloured Ehresmanoëdre.

Now, choose one colour (i+1, i) , and consider the monocolour subgraph $\Gamma_{i+1\ i}$ obtained by erasing the edges of colour different from the choosen one.

Then, writing i+1 i for the permutation 1 ... i-1 i+1 i i+2 ... , one has Monk's formula

$$2.2 \qquad X_{i+1\ i} \cdot\cdot X_v = \sum X_w$$

sum on all $w : \ell(w) = \ell(v) + 1$, vw is an edge of $\Gamma_{i+1\ i}$, i.e. there is an edge of colour (i+1 i) between v and w .

It is not too difficult to verify by induction this formula. The remarkable fact is that there is no multiplicity in this multiplication. Pieri's formula asserts that the multiplication of a Schur function by a special Schur function of any degree (i.e. elementary or complete symmetric functions, cf. Macdonald) produces no multiplicities. The same thing happens more generally for Schubert polynomials, i.e. the multiplication of a Schubert polynomial by a special one gives rise to no multiplicities (cf. L & S for the description of the w coming in the summation). Thus Monk's formula is the initial degree one-case of the general Pieri's formula.

As a by-product, one obtains a commutation property which is valid for all finite Coxeter groups.

Let C_{21}, C_{32} ... be the matrices of the directed graphs $\Gamma_{21}, \Gamma_{32}, \ldots$, i.e. we put 1 at the place (v,w) of the matrix $C_{i+1\ i}$ if $\ell(v) < \ell(w)$, and vw is an edge of $\Gamma_{i+1\ i}$, and 0 otherwise. Then one has

2.3 Lemma: *The matrices* $C_{i+1\ i}$ *commute.*

Proof.: As the multiplication of Schubert polynomials by $X_{i+1\ i}$ is described by the matrix $C_{i+1\ i}$. The lemma is equivalent to the fact that the product $X_{i+1\ i} \cdot X_{j+1\ j} \cdot X_v$ is equal to $X_{j+1\ j} \cdot X_{i+1\ i} \cdot X_v$ for every v.

This specific property of Bruhat order on Coxeter groups has to be proved in itself without reference to multiplication of Schubert polynomials.

§ 3 Cohomology of the flag manifold.

The reader who wants to use Figure 2 to multiply $a^3\left(=X_{4123}\right)$ by $a\left(=X_{2134}\right)$ finds no edge with colour 21 from 4123 upwards. So he must disagree with Monk's formula as stated above. But if he enlarges the alphabet by just one letter, he certainly obtains that

$$a^3 \cdot a = X_{41235} \cdot X_{21345} = X_{51234} = a^4 .$$

More generally, to use Monk's formula for the symmetric group W_n , one must for safety reasons imbed it into W_{n+1} .

Alternatively one can also notice that

$$a^4 = S_4(a+b+c+d) - (b+c+d)\,S_3(a+b+c+d) + (bc+bd+cd)\,S_2(a+b+c+d) - bcd(a+b+c+d)$$

which has the consequence that a^4 belongs to the ideal generated by the polynomials summetrical in a,b,c,d .

Definition: The cohomology ring of the flag manifold associated to the symmetric group W_{n+1} is

<u>3.1</u> $$\mathcal{H} = \mathbb{Z}[a,b,\ldots] \,/\, \mathcal{I}$$

where $\mathcal{I}$ is the ideal generated by the symmetric polynomials (with no constant term!) in all the variables (in other words, the ideal generated by the invariants of W_{n+1}).

It is easy to show by induction on n that H has two natural $\mathbb{Z}$-bases:

i) the monomials $a^I = a^{i_1} b^{i_2} \ldots$ with $0 \le i \le E$

ii) the Schubert polynomials X_w (the class of a Schubert polynomial in $\mathcal{H}$ is called a <u>Schubert cycle</u>).

Notice that $\mathcal{H}$ is of rank 1 in the maximal degree $\ell(\omega) = n(n+1)/2$ and that $X_\omega = a^E$.

Now, Monk's formula is perfectly valid: passing from an alphabet of $n+1$ letters to n , one annihilates exactly the Schubert polynomials X_w

FIGURE 3

MULTIPLICATION PAR a + b DANS L'ANNEAU DE COHOMOLOGIE

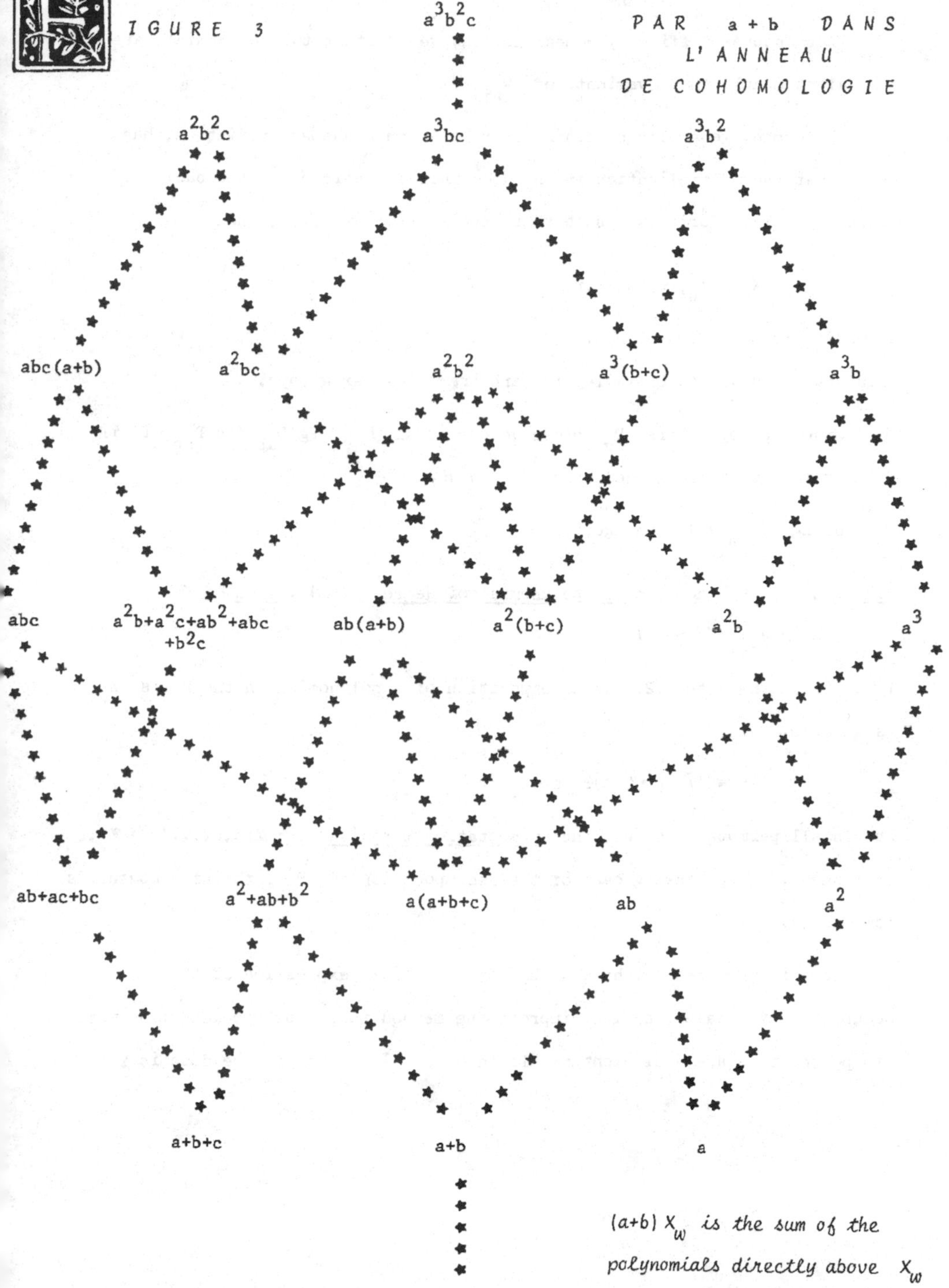

$(a+b)\,X_w$ *is the sum of the polynomials directly above* X_w

for those w such that $w_{n+1} \neq n+1$.

But another difficulty comes: how can we see that two polynomials are equivalent modulo the invariants of W_{n+1} ?

Ehresman, generalizing classical results on Grassmann varieties, has shown that the multiplication in H does induce a pairing on the basis of Schubert cycles: for w,w' such that $\ell(w) + \ell(w') = \ell(\omega)$, then

$$\underline{3.2} \qquad X_w \cdot X_{w'} \equiv a^E \text{ or } 0$$

according to $w' = \omega w$ or not.

This result is due to Chevalley for arbitrary Coxeter groups.

Since the operators D_w preserve the ideal I ($(fg)D_{ab} = g\, D_{ab} \cdot f$ if $f\, \sigma_{ab} = f$), they are indeed operators on H .

Using $a^E \partial_\omega = 1$, one gets:

3.3 Let P be a homogeneous polynomial of degree $\ell(\omega)$. Then

$P \equiv (P\, \partial_\omega)\, a^E \mod I$.

Thus, combining with 3.2, the decomposition of a polynomial in the basis X_w is given by

$$\underline{3.4} \qquad P \equiv \sum (P \cdot X_{\omega w})\partial_\omega\, \varepsilon \cdot X_w$$

sum on all permutations w , the augmentation morphism $\varepsilon : \mathbb{Z}[a,b,\ldots] \to \mathbb{Z}$, $a\varepsilon = b\varepsilon = \ldots = 0$ taking care of the decomposition of P into its homogeneous components.

Now, if one does not have at hand the explicit expression of the Schubert polynomials, one must improve the method to be able to determine when two polynomials are equivalent modulo the ideal I . This will be done in § 6.

§ 4 Projective degree of Schubert cycles.

Consider a graded ring H, call the graduation codimension, and assume that H is of rank 1 in maximal codimension (assumed different from infinity!) : $H^{max} \simeq \mathbb{Z}$.

Let Y in H be an element of codimension 1, and X of codimension d. Then the degree of X relative to Y is the image in $\mathbb{Z}$ of $X \cdot Y^{max-d}$.

When H is the cohomology ring of a projective variety, one chooses an imbedding in a projective space and Y is the class of the intersection with an hyperplane.

In our case, for the natural embedding of the flag variety, which is due to Plücker, Y is equal to the sum of all Schubert polynomials of codimension 1 :

4.1 $$Y = X_{2134\ldots} + X_{1324\ldots} + X_{1243\ldots} + \ldots = na + (n-1)b + (n-2)c + \ldots$$

(To distinguish between the degree of X relative to Y and the degree of X as a polynomial, we call the first projective degree and the second codimension.)

To compute the projective degrees, it is sufficient to know them for the Schubert cycles. In the case of grassmannians (a certain quotient of H) one obtains the degrees of the irreducible representations of the symmetric group, so these projective degrees should be interesting by themselves, regardless of their geometrical interpretations.

We have already done most of the work: as the multiplication by $X_{i+1\ i}$ corresponds to the edges of colour $(i+1\ i)$ in the coloured Ehresmanoëdre, one gets:

4.2 Proposition. The projective degree of X_w is the number of paths from w to ω in the coloured Ehresmanoëdre.

FIGURE 4

ORDRE D'EHRESMAN SUR LE GROUPE SYMETRIQUE

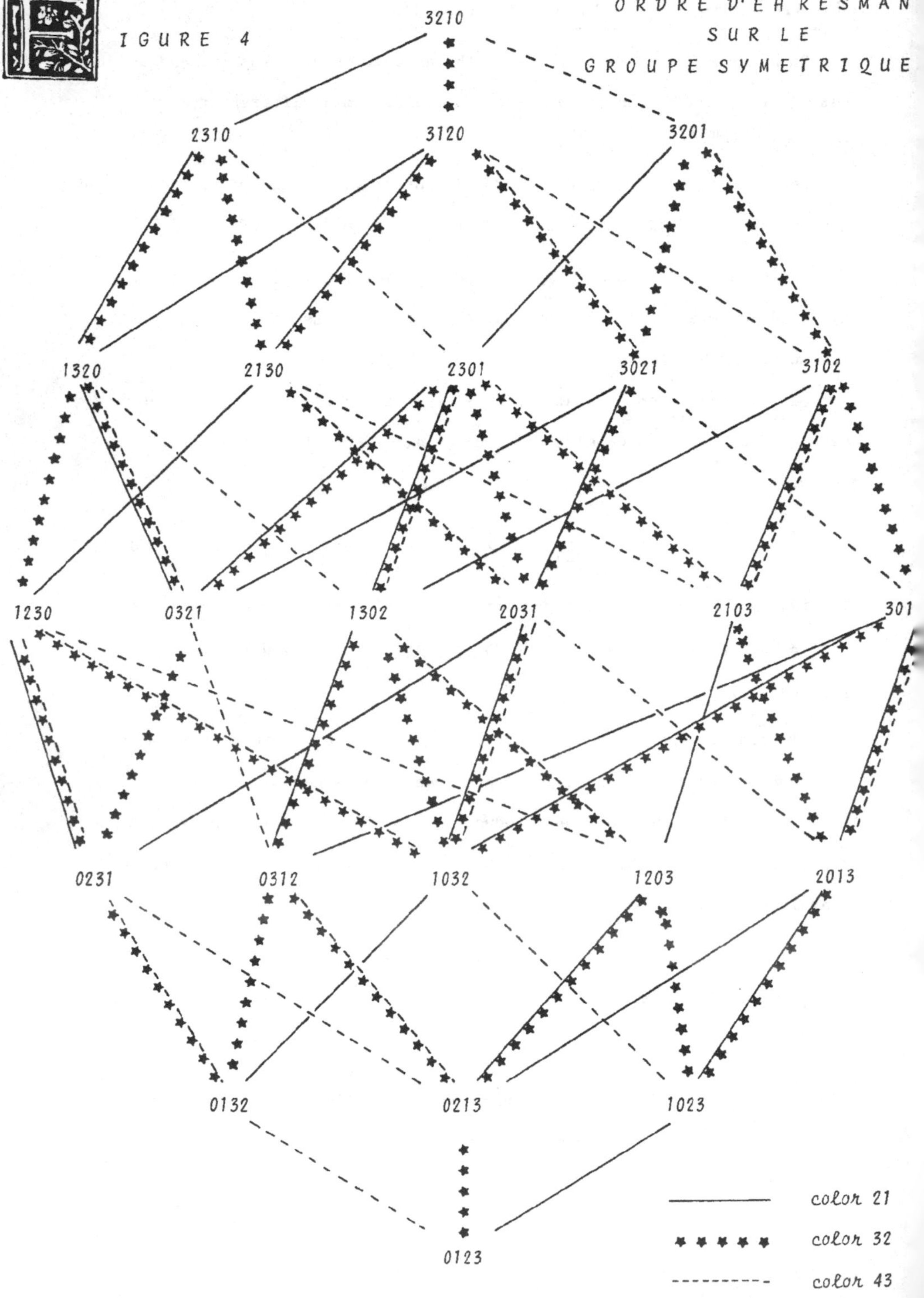

FIGURE 5

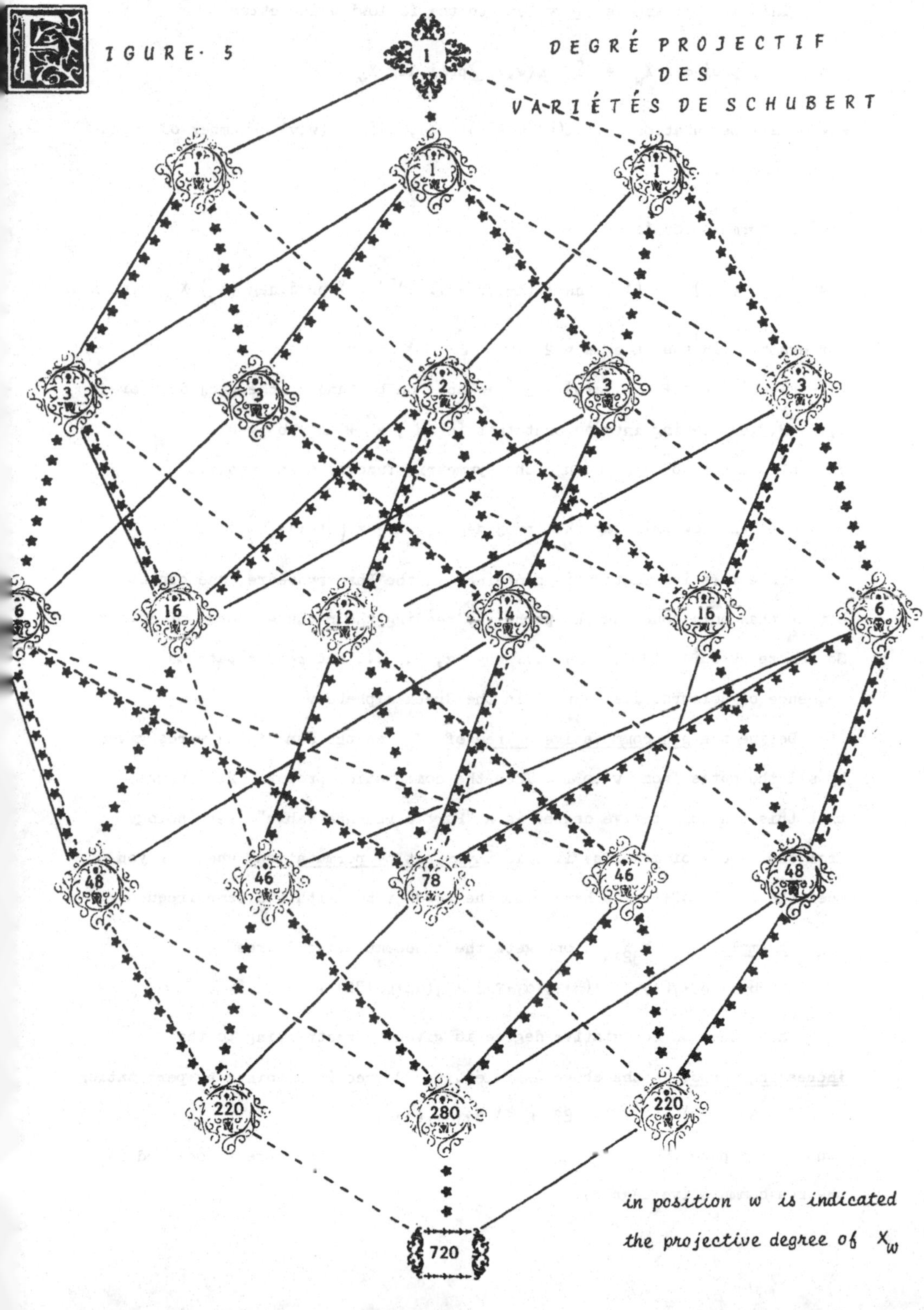

in position w is indicated the projective degree of X_w

This proposition is equivalent to the following induction:

4.3 $\quad \text{proj.deg } X_w = \sum_v m(w,v) \ \text{proj.deg } X_v$

sum on all permutations v: $\ell(v) = \ell(w) + 1$, with $m(w,v)$ = number of edges from w to v.

Another formulation is:

4.4 $\quad (1-Y)^{-1} = (1 - (na + (n-1)b + \ldots))^{-1} \equiv \sum \text{proj.deg } (X_w) \, X_{\omega w}$ in H.

For example, in the case $n = 2$
$(1 - (2a+b))^{-1} \equiv 1 + a + (a+b) + 3a^2 + 3ab + 6a^2b$ (and so the proj.deg are 1,1,1,3,3,6) taking into account that in H, $a^2 + ab + b^2 \equiv 0$, $a^3 \equiv b^3 \equiv a^2b + ab^2 \equiv 0$ modulo the symmetric functions in a,b,c.

One can show this way that proj.deg $X_{123\ldots} = \left(n(n+1)/2\right)!$

Since more information is contained in the Ehresmanoëdre, one can do better than only counting the paths, by reading the paths as words of colours. So denote colours (21), (32), ... by α, β, ..., and read a path as a sequence of colours, i.e. a word in the Greek alphabet.

Define the non-commutative degree of X_w as the sum of the words given by all the paths from w to ω. Then the commutation property 2.3 insures that this non commutative degree is a "Partie reconnaisable" (terminology from the theory of monoids) i.e. is invariant by permutation: whenever you meet the word $\alpha\alpha\beta\gamma$, you have also the word $\gamma\alpha\beta\alpha$ with the same frequency.

4.5 Example. For X_{3214}, one gets the noncommutative degree
$(\alpha+\beta+\gamma)(\beta\gamma+\gamma\beta+\gamma\gamma) + (\beta+\gamma)(\alpha\gamma+\gamma\alpha) + \gamma(\alpha\beta+\beta\alpha+\beta\beta)$.

Thus, the non commutative degree is given by restricting to the increasing words; in the above examples, the degree is obtained by permutation of (6) $\alpha\beta\gamma$ + (3) $\alpha\gamma\gamma$ + (3) $\beta\beta\gamma$ + (3) $\beta\gamma\gamma$ + (1) $\gamma\gamma\gamma$
(inside the parenthesis, we have indicated how many words are associated in the non-commutative degree).

In other words, if $\varphi : \mathbb{Z}\left[[\alpha,\beta,\ldots]\right] \to \mathbb{Z}[\alpha,\beta,\ldots]$ is the natural morphism (the evaluation) from the ring of non-commuting variables to the ring of polynomials, then the non-commutative degree is the inverse image of a polynomial $Z_{\omega,w}$ that we call the colour-degree of X_w.

More generally, one defines the polynomials $Z_{v,w}$, when $\ell(v) \geq \ell(w)$, to be the sum of all increasing paths from v to w (this will correspond to the degree of the intersection of two Schubert cycles); put $Z_{v,w} = 0$ if $\ell(v) < \ell(w)$.

If moreover, one defines M_α to be the matrix: the entry (v,w) is α or 0, according as $\ell(v) = \ell(w) + 1$ and there is an edge of colour α between v and w, or not, and similarly for $M_\beta, M_\gamma \ldots$, one obtains from 2.3 the commutation of the matrices $M_\alpha, M_\beta, \ldots$.

Exercise. Prove that $Z_{n+1\ 12..n,\ 12..n+1} = \sum \alpha^i \beta^j \gamma^k \ldots$, sum on all different monomials of total degree n, with $i \leq n$, $j \leq n-1$, $k \leq n-2$, ...

e.g. $Z_{4123,\ 1234} = \alpha(\alpha\alpha + \alpha\beta + \beta\beta + \alpha\gamma + \beta\gamma)$.

§ 5 The G-polynomials.

Instead of taking the cohomology ring of the flag manifold as we did in § 3, it is more fruitful to take another quotent of the ring of polynomials, which is called the Grothendieck ring of the flag manifold; denote the variables by $L_a, L_b, \ldots$ to distinguish from the preceeding case, and keep the same notations for the operators $\partial_{ab}, \pi_{ab}, \ldots$, as no ambiguity is to be feared.

Call θ the specialization ring-morphism: $L_a\theta = L_b\theta = \ldots = 1$ and let $\mathcal{J}$ be the ideal generated by the relations

5.1 $\forall\ f \in \mathbb{Z}[L_a,L_b,\ldots]$, $f\,\pi_\omega = f\,\pi_\omega\theta$

i.e. the totally symmetric polynomials are equalled to their value for $L_a = L_b = \ldots = 1$.

5.2 Definition. The Grothendieck ring of the flag manifold is

$$K = \mathbb{Z}[L_a, L_b, \ldots] / J .$$

The properties of this ring are strongly linked with those of symmetric functions, whose theory has been formalized in the theory of λ-rings.

As J is invariant under the action of the D_w , these operators still act on K . Of course, taking relations 5.1 instead of 3.1 do not change the $\mathbb{Z}$-bases of the quotient ring, so that one has

5.2 The set of monomials $L^I = L_a^{i_1} L_b^{i_2} \ldots$, for $0 \leq I \leq E$. is a $\mathbb{Z}$-basis of K .

5.3 The Schubert polynomials (in the alphabet $L_a, L_b, \ldots$) are a $\mathbb{Z}$-basis of K .

As $L_a L_b L_c \ldots \equiv 1$, we see that the ring K contains the inverse of the variables $L_a, L_b, \ldots$. The inversion of $L_a, L_b, \ldots$ extends to an involution morphism of K which is called duality by reference to vector bundles ($L_a, L_b, \ldots$ are the tautological line bundles of the flag manifold).

It is convenient to introduce new variables $x = 1 - L_a^{-1}$, $y = 1 - L_b^{-1}$, $z = 1 - L_c^{-1}$ .... The symmetrizers associated to $x, y, \ldots$ are related to those of $L_a, L_b, \ldots$; One checks

5.4 $$\forall f \in K , \quad f \pi_{ab} = (f - fy)\partial_{xy}$$

and similarly for all pairs of consecutive letters, which, incidentally, shows that a change of variables induces a non trivial transformation of the symmetrizers. As in § 2, we choose a maximal element

$$G_\omega = x^E = x^n y^{n-1} z^{n-2} \ldots$$

and we define the G-polynomial indexed by the permutation w by

5.5 $$G_w = G_\omega \, \pi_{\omega w} \; .$$

Figure 6 gives the case $n = 3$.

It is clear from lemma 5.4 that the homogeneous part of smallest degree (= $\ell(w)$) of G_w is the Schubert polynomial X_w (in the variables $x,y,\ldots$ instead of $a,b,\ldots$). Thus the Schubert polynomials are nothing but the leading term of the G_w.

Schubert polynomials (in $x,y,\ldots$) being a basis of K, one can express the G_w in term of the X_w, or conversely, the X_w in term of the G_w, the matrix being triangular.

e.g. for $(w) = 4$, one has

$$X_{2413} = G_{2413} + G_{3412} \; ; \qquad X_{2341} = G_{2341} \; ;$$

$$X_{3142} = G_{3142} + G_{3241} \; ; \qquad X_{3214} = G_{3214} \; ;$$

$$X_{1432} = G_{1432} + 2G_{2431} + G_{3421} + G_{3412} \; ; \qquad X_{4123} = G_{4123} \; .$$

To express the multiplication in the basis of G_w is more complicated than in the basis X_w. We shall give elsewhere the corresponding "Pieri-formula". For example, one reads on figure 5 that

$$G_{1324} \, G_{1324} = (x+y - xy)^2 = G_{2314} + G_{1423} - G_{2413} \; .$$

(We previously had $X_{1324} \cdot X_{1324} = X_{2314} + X_{1423}$; here we had to substract the supremum of 2314 and 1423 which is 2413 ; bigger intervals are involved in general.)

To understand the link between the two rings H and K, one must recall the existence and properties of Chern classes:

Denote by $1 + H^+$ the multiplicative monoid of polynomials with constant term 1 (and coefficients in $\mathbb{Q}$). There exists a homomorphism:

$$c : K \to 1 + H^+$$

FIGURE 6

ANNEAU DE GROTHENDIECK
DES
VARIÉTÉS DE DRAPEAUX

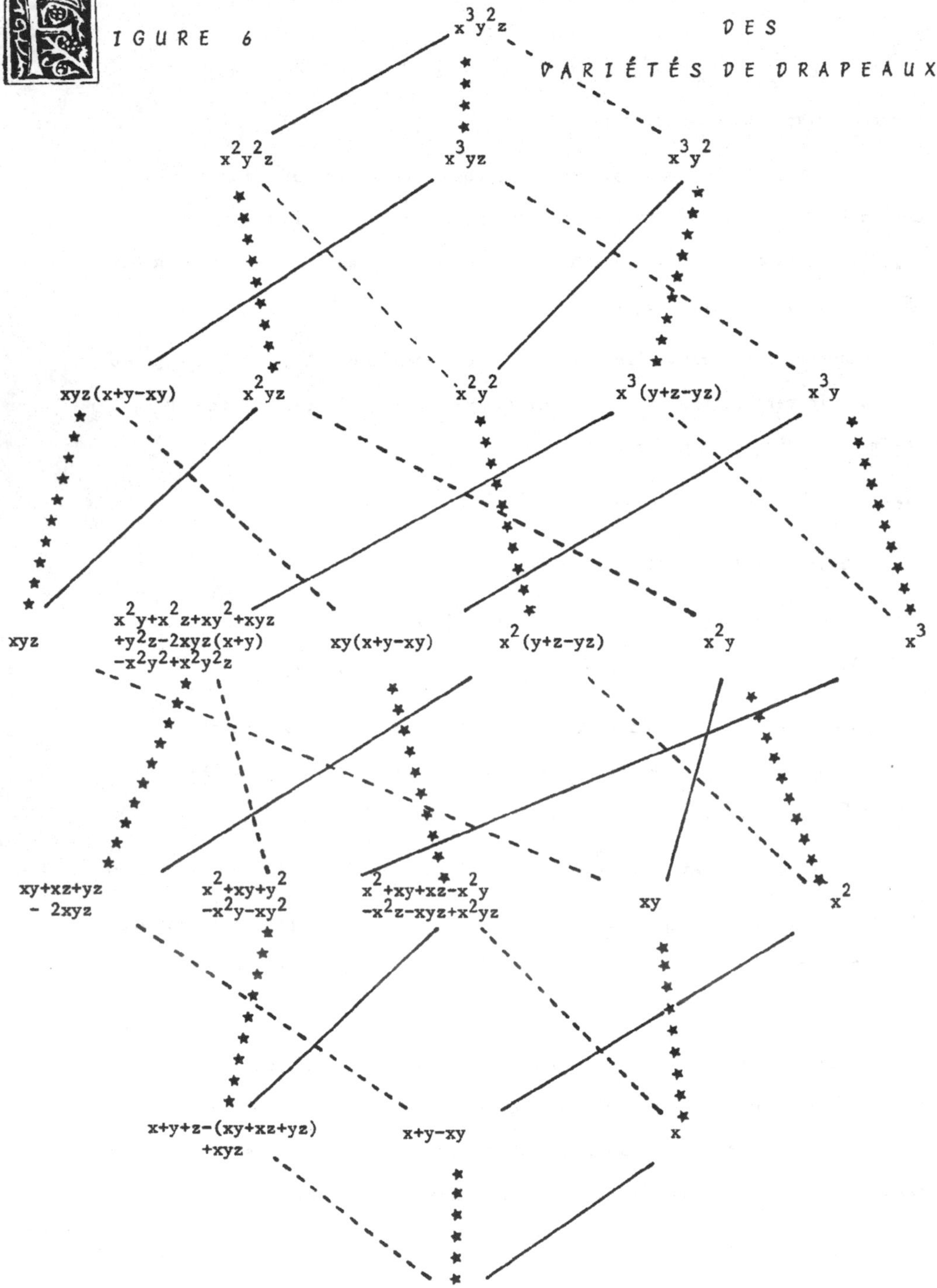

such that $c(1) = 1$, $c(E+F) = c(E)\cdot c(F)$. On the basis L^I , it takes the values

$$c(L_a^{\,i}\, L_b^{\,j}\, L_c^{\,k} \ldots) = 1 + ia + jb + kc + \ldots .$$

(Of course the multiplication in K induces a "product" in $1 + H^+$, for whose explicit description one needs Schur functions - see Macdonald.)

Now one can check

$$c(G_w) = 1 - (-1)^{\ell(w)} \; (\ell(w)-1)! \;\; X_w + X'$$

where X' is a polynomial of degree $> \ell(w)$. (cf. SGA6, exp. 0 formule 1.18).

If G is a sum $\Sigma\, n_w G_w$ with w of constant length $\ell(w) = d$ (one says $G \in K^d$)

$$c(G) = 1 - (-1)^d \;\; (d-1)! \;\; \Sigma\, n_w X_w + X'$$

and

$$c(G\pi_{ab}) = 1 - (-1)^{d-1}\, (d-2)! \;\; \Sigma\, n_w X_w \partial_{ab} \;+\; X'\partial_{ab}$$

so that one sees that:

<u>5.6</u> <u>Proposition</u>: $-(d-1/^{-1}\,\partial_{ab}$ <u>is the image by the Chern homomorphism of</u> π_{ab} <u>acting on</u> K^d .

§ 6 Quadratic form on the cohomology ring.

Most of the preceeding description relies heavily upon the natural bases of the cohomology or Grothendieck ring of the flag variety. To be able to compute without restriction in these rings, one must be able to express a general element (in a finite time) in the bases already defined.

The operators corresponding to the permutation of greatest length are the most effective rool for this purpose. It amounts to define on each of the spaces H and K a quadratic form.

We consider sequences as vectors in $\mathbb{Z}^{n+1}$ and thus can write $I \pm J$; through the identification $I \leftrightarrow a^I$, the symmetric group acts on sequences: $I \rightsquigarrow Iw$;

recall that E is the sequence $n, n-1, \ldots, 0$.

Now, when $-E \leq I \leq E\omega$, one checks from formula 1.9 that

$$\underline{6.1} \qquad a^I \pi_\omega = a^{I+E} \partial_\omega \begin{cases} = (-1)^{\ell(w)} & \text{if there exists } w \text{ such that } I + E = Ew \\ = 0 & \text{otherwise.} \end{cases}$$

Moreover, $a^I \pi_\omega = (a^{-I}\omega)\pi_\omega$, and $a^I \pi_\omega = 0$ if $i_1 + \ldots + i_{n+1} (=|I|) \neq 0$.

E.g. $I = -3102 \Rightarrow I+E = 0312$ is a permutation of E and so is $-I\omega + E$ $(= -2\ 0\ -1\ 3 + 3\ 2\ 1\ 0 = 1\ 2\ 0\ 3)$.

Owing to this symmetry between I and $-I\omega$, one defines a scalar product on H by its values on the basis a^I (for $0 \leq I \leq E$) :

$$\underline{6.2} \qquad (a^I, a^J) = a^{I\omega - J}\, \pi_\omega .$$

For example, for four letters and degree 3

2100	1110	2010	1200	3000	0210	I \ J
0	0	0	0	0	−1	2100
0	0	0	0	−1	0	1110
0	0	0	−1	1	1	1200
0	0	−1	0	1	1	2010
0	−1	1	1	−1	0	3000
−1	0	1	1	0	−1	0210

On the example one sees that the quadratic form, for a given weight, is positive or negative definite, and triangular for an appropriate ordering of the monomials.

Instead of describing the ordering, which directly comes from the interpretation of sequences I such that $0 \leq I \leq E$ as coding permutations, one can do better and give the adjoint basis of a^I.

<u>6.3</u> Let $P_I = \Pi_{0 \leq p \leq n} \Lambda_{i_p}(A_{n-p})$.

A_p being the alphabet of the first p letters, and Λ_i the elementary symmetric function of degree i ; then one has

<u>6.4</u> <u>Proposition: The family</u> $(-1)^{|I|}\{P_I\}$ <u>with</u> $0 \leq I \leq E$, <u>is the adjoint basis of</u> $\{a^I\}$.

For example, for $n=3$, $I = 2010$, $P_I = \Lambda_2(a+b+c)\,\Lambda_1(a) = a^2b + a^2c + abc$, and one checks from the preceeding table that

$$(a^I, P_I) = -1 \ , \quad (a^J, P_I) = 0 \quad \text{if} \quad J \neq I .$$

<u>6.5</u> <u>Corollary</u> (<u>Bott-Rota's straightening</u>).

<u>If</u> Y <u>is an homogeneous polynomial of degree</u> d, <u>then in</u> H, $(-1)^d\, Y \equiv \Sigma\,(Y, P_I)\, a^I \equiv \Sigma\,(Y, a^I)\, P_I$ <u>sum on all sequences</u> I <u>such that</u> $0 \leq I \leq E$.

Thanks to 6.1, this straightening is an efficient way of decomposing in H the class of a polynomial. For the decomposition in the basis X_w, we already have 3.4.

We note that $X_w\, \partial_v = 0$ if $\ell(v) = \ell(w)$ and $v \neq w^{-1}$, because $X_w\, \partial_v = X_\omega\, \partial_{\omega w}\, \partial_v$, so that either $\omega w v = \omega$, or $\ell(\omega w v) < \ell(\omega) \Leftrightarrow \partial_{\omega w}\, \partial_v$ can be written $\partial_{w'} \partial_u\, \partial_u\, \partial_{w''}$ $(= 0)$. Thus we have the other way of decomposing a polynomial Y.

<u>6.6</u> $$Y \equiv \Sigma\ (Y \cdot \partial_{w^{-1}})\,\varepsilon\, X_w$$

sum on all permutations w, which $P\varepsilon$ = term of degree o of P, as in 3.4.

§ 7 Quadratic form on the Grothendieck ring.

As Schubert polynomials are still a basis of the Grothendieck ring K, one could still keep the scalar product for which the Schubert polynomials are an orthonormal basis. This would not fit well with the action of the operators on K.

Remembering that π_ω sends K to its subring $\mathbb{Z}$, one can define

7.1 $\quad \forall\ P, Q \in K\ ,\ \langle P, Q \rangle = (PQ)\,\pi_\omega\ ;$

on the basis G_w, the quadratic form takes only the values 0 or 1.

For example, for S_3, the multiplication table is

	G_{123}	G_{213}	G_{132}	G_{231}	G_{312}	G_{321}
G_{123}	G_{123}	G_{213}	G_{132}	G_{231}	G_{312}	G_{321}
G_{213}	G_{213}	G_{312}	$G_{312}+G_{231}-G_{321}$	G_{321}	0	0
G_{132}	G_{132}	$G_{312}+G_{231}-G_{321}$	G_{231}	0	G_{321}	0
G_{231}	G_{231}	G_{321}	0	0	0	0
G_{312}	G_{312}	0	G_{321}	0	0	0
G_{321}	G_{321}	0	0	0	0	0

and the quadratic form is the image of this table by π_ω (noting that $\forall w\ ,\ G_w\,\pi_\omega \equiv 1$):

1	1	1	1	1	1
1	1	1	1	0	0
1	1	1	0	1	0
1	1	0	0	0	0
1	0	1	0	0	0
1	0	0	0	0	0 .

We shall not prove here the two following propositions which generalize 3.4 and 6.6.

7.2 Proposition. For any w , let H_w be the sum of G-polynomials $\Sigma_{v \geq w} \ (-1)^{\ell(v)-\ell(w)} \ G_v$.

Then $\{H_{\omega w}\}$ is the adjoint basis of $\{G_w\}$ (with respect to $\langle \, , \, \rangle$).

For example,

$$\langle G_{132} - G_{231} - G_{312} + G_{321} \, , \, G_w \rangle = 0 \quad \text{except for} \quad w = 312 = \omega \cdot 132 \, .$$

7.3 Proposition. Let θ be the specialization morphism $L_a \theta = L_b \theta = \ldots = 1$. Then $\forall \, P \in K$, $\forall w$, $\langle Y, G_w \rangle = Y \, \pi_{w^{-1}} \, \theta$.

As for every w , $\langle G_w, 1 \rangle = (G_w \cdot 1) \, \pi_\omega = 1$ the fact that $\langle H_w, 1 \rangle = 0$ generalizes the property of the Moebius function (for the Bruhat order) to be ± 1 .

§ 8 Applications.

We have mainly described the tools to study the cohomology or Grothendieck ring of the flag manifold. Many questions arising from the theory of groups or algebraic geometry can be then easily studied.

8.1 The representation of the symmetric group W_{n+1} on H or K . One must note that as $\mathbb{Z}$-modules, H and K are isomorphic to the regular representation of W_{n+1} but that the degree gives us an extra information; in fact, the multiplicity of an irreducible representation can be considered as a polynomial (which happens to be a Kostka-Foulkes polynomial coming in the theory of representation of the finite linear groups). More generally, De Concini et Procesi have studied the quotients of H associated to the variety of flags fixed by a given unipotent matrix.

8.2 Enumerative geometry on the flag manifold.

We have only given the projective degree of a Schubert cycle in § 4.

One needs also the postulation of the cycle X_w with respect to a line bundle L^I : by definition, it is $\Sigma\,(-1)^i \dim H^i\,(\mathcal{O}_w, L^I)$; once given the rules of the translation, it simply becomes $(L^I G_w)\,\pi_\omega$, which is also equal to $L^I\,\pi_{w^{-1}}\,\theta$, as asserted in 7.3.

The Chern classes of a variety are the first invariants of it that one tries to get. In the case of the flag manifold, the tangent bundle T has class

$$L_a \cdot L_b^{-1} + L_a \cdot L_c^{-1} + L_b L_c^{-1} + \ldots \quad \text{in } K$$

so that its Chern class is

$$c(T) = (1+a-b)\,(1+a-c)(1+b-c)\ \ldots$$

and it remains to compute $c(T)$ in the basis of Schubert cycles. This will be done elsewhere.

8.3 Representations of the linear group $G\ell(\mathbb{C}^{n+1})$.

One can consider the ring of invariants of W_{n+1}: $\mathbb{Z}[a,1/a,b,1/b,\ldots]^W$ to be the ring of formal sums of representations of $G\ell(\mathbb{C}^{n+1})$.

Bott's theorem evaluates in this ring, for any line bundle L^I, $I \in \mathbb{Z}^{n+1}$, and any i , the representation $H^i(X, L^I)$, X being the flag manifold.

We have obtained here a little less:

$$\Sigma\,(-1)^i\, H^i(X, L^I) = (L^I)\,\pi_\omega\ ,$$

(in fact, all the $H^i(X, L^I)$ are $\{0\}$ except at most one, so that the two computations are not very different).

One can also look for syzygies of the Schubert variety corresponding to w , i.e. try to get a complex of locally free bundles which "solves" the ring of the Schubert variety. The class of the complex in $\mathbb{Z}[a,1/a,b,1/b,\ldots]$ is given by

$$[\,(1-L_a^{-1})^n\,(1-L_b)^{n-1}\ \ldots]\ \pi_{\omega w}$$

but, of course there remains to describe the morphisms inside the complex. This, we shall not do.

8.4 Root systems and Coxeter groups.

Most of the properties of the operators D_w can be extended to other finite Coxeter groups, as shown first by Demazure and independently Bernstein, Gelfand-Gelfand.

In the case of the symmetric group, if $\alpha, \beta, \ldots,$ are the simple roots, and ρ half the sum of positive roots, then $e^{\alpha}, e^{\beta}, \ldots$ are respectively $L_aL_b^{-1}, L_bL_c^{-1}, \ldots,$ and e^{ρ} is equal to L^E up to a power of $L_aL_bL_c \ldots$. If I is weakly decreasing (L^I is dominant), then Weyl's character formula for the corresponding irreducible representation E_I is

$$ch(E_I) = \Sigma(-1)^w (L^I L^E)w \,/\, \Sigma(-1)^w L^E w$$

and as we have remarqued in 1.9, it can be written

$$ch(E_I) = (L^I L^E)\,\partial_\omega \quad ;$$

an equivalent result, using instead the operator π_ω , is

$$ch(E_I) = L^I\,\pi_\omega$$

which is in fact Bott's formula.

8.5 Determinants.

For $m : 0 \leq m \leq n+1$, one can consider The Grassmann variety of subvector spaces of dim $m+1$ of $\mathbb{C}^{n+1}$. The associated cohomology ring is the subring $H^{W' \times W''}$ of H invariant under the product of symmetric groups $W' \times W''$ (W' being the group on the first $m+1$ letters, W'', on the remaining letters).

A $\mathbb{Z}$-basis of $H^{W' \times W''}$ is the set of Schubert polynomials X_w for the w of minimum length in their class modulo $W' \times W''$. In this case X_w is a Schur function on the alphabet of the first $m+1$ letters, and all the properties of the cohomology ring of the Grassmann variety (or of the Grothendieck ring) can be translated in term of Schur functions. In particular, the determinantal expression of Schur function gives rise to a determinantal expression for X_w (due to Giambelli), for G_w , for the postulation (due to Hodge), etc...

Unfortunately, not all permutations in general give determinants. We have given in [L & S] several characterizations of those permutations for which X_w, G_w, ... are determinants (permutations vexillaires); for them, the computations are very similar to the ones in the more special case of Grassmannvariety.

8.6 Combinatorics.

A combinatorial and powerful description of Schur functions is given by Ferrers diagrams and Young tableaux. One can similarly associate to any permutation a diagram (due to Riguet), and fill it according to rules deduced from Pieri-Monk's formula 2.2. This seems to be an interesting generalization of Young tableaux and the plactic monoid.

BERNSTEIN I.N., GELFAND I.M. and GELFAND S.I., *Russian Math.Surv.* 28, 1973, p.1-26.

DEMAZURE M., *Inv. Math.* 21, 1973, p. 287-301.

EHRESMAN C., *Annals of Maths.* 35, 1934, p. 396-443.

HILLER H., *Geometry of Coxeter groups* (Pitman, 1982).

LASCOUX A. and SCHÜTZENBERGER M.P., C.R. Acad. Sc. Paris 294, 1982, p. 447.

MACDONALD I.G., *Symmetric and Hall polynomials* (Oxford Math. Mono., 1979).

ON SOME RESTRICTION THEOREMS FOR SEMISTABLE BUNDLES

Vikram B. Mehta

Department of Mathematics
University of Bombay
Bombay 400 098

Introduction

In this survey article, we shall mainly be concerned with the following problem: given a semistable vector bundle on a smooth projective variety X, describe its restriction to a general hyperplane section X. In characteristic zero, one can prove that the instability degree of the restriction is bounded by the degree of X (Thm. 2.1). The characteristic zero assumption is used to prove that a certain morphism is constant if its differential is zero. This is precisely what fails in char p, where we have to factor the given morphism by its Frobenius transform [2].

The use of the "standard construction" was used by Van de Ven in studying uniform bundles, and then by Grauert-Mullich for rank 2 bundles on $\mathbb{P}^2$ and then by Forster--Hirschowitz-Schneider and Maruyama for bundles of arbitrary rank on any variety.

Another method which can be used is to restrict the bundle to hypersurfaces of high degree [11]. This enables one to prove that the restriction to a hypersurface of sufficiently high degree is semistable. We shall also sketch a proof of a theorem of Maruyama [9] where he proves that if $\text{rk}V < \dim X$, then the restriction of V to any hyperplane section of X is again semistable. This enables us among other things, to prove immediately that the set of semistable bundles on X of rank 2 and fixed Chern Classes is bounded, for arbitrary X.

We shall not touch upon the topics of uniform bundles on $\mathbb{P}^n$, for which we refer the reader to [4] for the classification in char. 0 and to [2,6] for characteristic $p > 0$. Also we shall not mention the restriction theorems proved for special bundles on $\mathbb{P}^3$, such as the results of Barth [1] on null-correlation bundles in Char. 0 and those of Ein [2] in Char. $p > 0$. The reader is also urged to consult [3] which appeared too late for inclusion in this survey.

Throughout this paper, we use the notions of semistability and stability in the sense of Mumford-Takemoto [9, 11]. For the sake of convenience we recall these notions as well as those relating to the Harder-Narasimhan forunstable filtration of a vector bundle on a variety [7, 9].

Definition 0.1: Let X be smooth projective of $\dim n$ and $\theta_X(1)$ be very ample on X. For a coherent torsion-free sheaf V on X, define $\deg V = c_1(V)\cdot$ $\cdot\, \theta_X(1)^{n-1}$ and $\mu(V) = \deg V/\mathrm{rk} V$. Define V stable (resp. semistable) if for all $W \subset V$, we have $\mu(w) < \mu(v)$ (resp. $\leq \mu(V)$).

Suppose V is not semistable. Then there exists a subsheaf V_1 of V with $\mu(V_1) = \sup\limits_{W \subset V} \{\mu(w)\} = \mu_o$ say, and V_1 having maximal rank among subsheaves W of V with $\mu(W) = \mu_o$.
Further V_1 is unique and infinitesimally unique, i.e.
$\mathrm{Hom}(V_1, V/V_1) = 0$ [7]. Now look at V/V_1. Then there exists a subsheaf V_2 of V such that $V_1 \subset V_2 \subset V$ and V_2/V_1 is the maximal subsheaf, in the above sense, of V/V_1. Continue this process to get a flag

$$0 = V_o \subset V_1 \subset \ldots \subset V_r = V$$

with the properties:

1) Each V_i/V_{i-1} is semistable, $1 \leq i \leq r$.
2) $\mu(V_i/V_{i-1}) > \mu(V_{i+1}/V_i)$, $\quad 1 \leq i \leq r-1$.
3) The above flag is unique.

We put $\mu_i(V) = \mu(V_i/V_{i-1})$ and we call V_1 the β-subsheaf of V. We also $\mu_{max}(V) = \mu_1$ and $\mu_{min}(V) = \mu_r$. The above filtration (1) is called the Harder--Narasimhan filtration of V. By the uniquenes of the flag, it is defined over any field of definition of V.

Let $f : X \to S$ be a smooth projective family of varieties and V a vector bundle on X such that $V_K|X_K$ is unstable, where K is the quotient field of S. Then there exists a nonempty open subset U of S and a filtration

$$0 = W_o \subset W_1 \subset \ldots \subset W_r = V/f^{-1}(U)$$

such that for every point $x \in U$, $0 = W_{o,x} \subset W_{1,x} \subset W_{r,x}$ is the unstable filtration of $V/f^{-1}(x)$.

Section I

First we introduce some notation. Let X be a smooth projective variety of dimension n. Let $(\alpha_1,\ldots,\alpha_{n-1})$ be a sequence of integers with each $\alpha_i \geq 2$. For any $m > 0$, put $\underline{m} = (\alpha_1^m, \ldots, \alpha_{n-1}^m)$. Put $S_{\underline{m}}$ = the multi-projective space of homogeneous polynomials of multi-degree $\alpha_1^m, \ldots, \alpha_{n-1}^m$, let $Z_{\underline{m}}$ be the correspondence variety:

$$\begin{array}{ccc} Z_{\underline{m}} & \xrightarrow{q_{\underline{m}}} & S_{\underline{m}} \\ {\scriptstyle p_{\underline{m}}}\downarrow & & \\ X & & \end{array}$$

For a closed point $x \in S_{\underline{m}}$, we call $q_{\underline{m}}^{-1}(x)$ a curve of type $\underline{m}$.
Let $K_{\underline{m}}$ be the quotient field of $S_{\underline{m}}$ and we define $Y_{\underline{m}}$, the generic complete intersection curve of type $\underline{m}$, by the fibre product:

$$\begin{array}{ccc} Y_{\underline{m}} & \longrightarrow & K_{\underline{m}} \\ \downarrow & & \downarrow \\ Z_{\underline{m}} & \longrightarrow & S_{\underline{m}} \end{array}$$

Now we can state

Theorem 1.1 [11]: Let V be a semistable vector bundle on X. Then $\exists\ m_0$ s.t. $V/Y_{\underline{m}_0}$ is semistable.

We sketch a proof. First assume that $\mathrm{rk}V = 2$, $\dim X = 2$ and $\mathrm{Pic}\ X = \mathbb{Z}$. Suppose $V/Y_{\underline{m}}$ is not semistable $\forall\ m > 0$. Let $L_{\underline{m}} \in \mathrm{Pic}(Y_{\underline{m}})$ contradict the semistability of $V/Y_{\underline{m}}$. Then $M_{\underline{m}} \in \mathrm{Pic}(X)$ such that $M_{\underline{m}}/Y_{\underline{m}} = L_{\underline{m}}$.

This is shown by taking a Lefschetz pencil of hyperplane sections of X for $m >> 0$ and then deducing the existence of an imbedded curve in X of type $\underline{m}$ which is integral and singular. Or, one can appeal to a theorem of Weil [13].

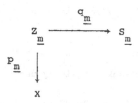

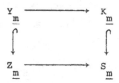

Next, we take a ruve $C_{\underline{m+1}}$ of type $\underline{m+1}$ and degenerate it, over a discrete valuation ring A, to α_1 curves $C_{\underline{m}}^{(i)}$ of type $\underline{m}$, $1 < i < \alpha_1$. Let $D \to A$ be the family and extend $L_{\underline{m+1}}$, which is defined on the general fibre of D, to a line bundle L_A on D and which is contained in V/D. Restrict L_A to the components of the special fibre. By comparing degrees, we get the important inequality:

$$\deg L_{\underline{m+1}} \leq \alpha_1 \deg L_{\underline{m}} \quad . \tag{1}$$

This shows that $\{\deg M_{\underline{m}}\}$ is bounded above. Since $\mathrm{Pic}\, X = \mathbb{Z}$, there exists a unique $M \in \mathrm{Pic}\, X$ such that $M/Y_{\underline{m}}$ is the β-subbundle of $V/Y_{\underline{m}}$ for all $m >> 0$. By Enriques-Severi, the inclusion $M/Y_{\underline{m}} \subset V/Y_{\underline{m}}$ lifts to an inclusion $M \subset V$ on X, contradicting the semistability of V on X.

Now assume $\mathrm{rk}V$, $\dim X$ are arbitrary. Let $W_{\underline{m}}$ be the β-subbundle of $V/Y_{\underline{m}}$, which may be assumed to be of constant rank. Again by degenerating $C_{\underline{m+1}}$ to $\alpha = \prod_{i=1}^{n-1} \alpha_i$ curves of type $\underline{m}$, we get an analogous inequality:

$$\deg W_{\underline{m+1}} \leq \alpha \deg W_{\underline{m}} \tag{2}$$

Again this shows the existence of a unique $M \in \mathrm{Pic}(X)$ such that $M/Y_{\underline{m}} = \det W_{\underline{m}}$ for all $m >> 0$. Now one can construct, for each x in $S_{\underline{m}}$ for $m >> 0$, a subsheaf W of V on X such that $W/q_{\underline{m}}^{-1}(x) = W_{\underline{m}}/q_{\underline{m}}^{-1}(x)$. Hence $\exists$ $\underline{m}_o$ such that $V/Y_{\underline{m}_o}$ is semistable.

By using inequality (2) one can prove that $V/Y_{\underline{m}}$ is semistable $\forall$ $m \geq m_o$. So if V_1 and V_2 are two semistable bundles on X, then both V_1 and V_2 are semistable on $Y_{\underline{m}}$ for some m. In particular, if we are in characteristic zero, then $V_1 \otimes V_2$ is also semistable on $Y_{\underline{m}}$, and hence also on X. This has also been proved by Maruyama and Ramanathan-Ramanan ([8], [12]).

Section II

In this Section we consider the following situation: Let $X \subset \mathbb{P}^n(\mathbb{C})$ be a smooth projective variety, $\dim X = r$. Let V be a semistable vector bundle on X. Put $t = n-r+1$ and let $G = G(t,n)$ be the Grassmann variety of t-planes in $\mathbb{P}^n$. Consider the following diagram

$$\begin{array}{ccc} Z & \xrightarrow{\;g\;} & G \\ {\scriptstyle f}\downarrow & & \\ X & & \end{array}$$

where Z is the correspondence variety, $Z = \{(x,W) \in X \times G / x \in W\}$. We consider $f^*(V)$ as a family of vector bundles on the fibres of g. We may assume $\exists$ an open set U of G and a filtration

$$0 = W_o \subset W_1 \subset \dots \subset W_s = f(V)/g^{-1}(U)$$

such that for all $x \in U$,

$$0 = W_{o,x} \subset W_{1,x} \subset \dots \subset W_{s,x} = V/g^{-1}(x)$$

is the Harder-Narasimhan filtration for $V/g^{-1}(x)$.

Now we have the following, which is due independently to Forster-Hirschowitz--Schneider and Maruyama [5,8]:

Theorem 2.1: We have $\mu(W_i) - \mu(W_{i+1}) \le d =$ degree X, $1 \le i \le s-1$.

Proof: Suppose i such that $\mu(W_i) - \mu(W_{i+1}) > d$. If rank $W_i = p$, W_i gives us a map $\sigma : g^{-1}(U) \longrightarrow G_p[f^*(V)|g^{-1}(U)] = H$ say. Let $h : H \to X$. Denote by T_Z the tangent bundle of Z and by T_f and T_h the relative tangent bundles of f and h respectively. The map σ defines $d\sigma : T_z \twoheadrightarrow T_h$. We restrict $d\sigma$ to $g^{-1}(x)$, $x \in U$. Now the restriction of $T_f/g^{-1}(x)$ can be identified with $\Omega^1(1)^{\oplus n-t}/g^{-1}(x)$, where $\Omega^1(1)$ is the bundle of twisted 1-forms on the t-space corresponding to x. Denote $T_f/g^{-1}(x)$ by B. Now $\mu_{min}(B) \ge -d$ and $\mu_{max}(B) \le 0$. And $T_h/g^{-1}(x) = (W_i \otimes V/W_i)$, which, by assumption on the μ_i's is seen to be filtered by semistable bundles whose μ's are $< -d$. (Here, we use the fact that tensor products of semistable bundles over $\mathbb{C}$ is semistable). Hence $d\sigma : T_f \longrightarrow T_h$ is zero for every $x \in U$. Thus σ provides us with a map $\tilde{\sigma}: X \to H$ and hence W_i descends to a subsheaf of V on X, contradicting the semistability of V.

Remark 2.2: [8] Assume in the above theorem $X = \mathbb{P}^n$ and we restrict a semistable V on X to a general $\mathbb{P}^r \subset \mathbb{P}^n$. Then if $0 = V_o \subset V_1 \subset \dots \subset V_s = V/\mathbb{P}^r$ is the Harder-Narasimhan filtration for $V/\mathbb{P}^r$, we have $\mu_i(V) - \mu_{i+1}(V) \le \frac{1}{r}$, $1 \le i \le s$. For the proof, we note that if $G = G(r,n)$, and $x \in G(r,n)$ then $T_f/G^{-1}(x)$ is semi-stable and $\mu[T_f/g^{-1}(x)] = \frac{-1}{r}$. Hence if $\mu_i(V) - \mu_{i+1}(V) > \frac{1}{r}$ for some i, V_i descends to a subsheaf of V, contradicting the semistability of V.

Remark 2.3: Assume that X is an r-dimensional complete intersection in $\mathbb{P}^n$, of multi-degree $(d_1,\ldots,d_{n-r})$. Let $d = \deg X = \Pi\, d_i$. Then for any semistable V on X and for any general $\mathbb{P}^t \subset \mathbb{P}^n$ with $t+r \geq n+1$, if $0 = V_o \subset V_1 \ldots \subset V_s = V/\mathbb{P}^t \cdot X$ is the Harder-Narasimhan filtration of $V/\mathbb{P}^t \cdot X$, we have $\mu_i(V) - \mu_{i+1}(V) \leq \frac{d}{t}$. The proof follows from Remark 2.1 and (1) of Section I. One has to show that $\mu_{\min}[T_f/\mathbb{P}^t \cdot X] \geq d\, \mu_{\min}[T_f/\mathbb{P}^t \cdot \mathbb{P}^r]$.

Remark 2.4: In [15] Schneider has proved that if V is a rank 3 semistable bundle on $\mathbb{P}^3(\mathbb{C})$ with $\deg V = 0$ then V restricted to a general $\mathbb{P}^2$ is again semistable. Making use of Maruyama's theorem [9, thm. 3.1] and 2.2 above, we can generalize this as follow:

Let V be rank n semistable on $\mathbb{P}^n(\mathbb{C})$ with $\deg V \neq -1$ or $1-n$. Then V restricted to a general $\mathbb{P}^{n-1}$ is again semistable. For, if the restriction is not semistable, then by [10], the Harder-Narasimhan filtration of $V/\mathbb{P}^{n-1}$ is

$$0 \to W_1 \subset V/\mathbb{P}^{n-1} \to W_2 \to 0$$

with rank $W_1 = 1$ or $n-1$. Further, we must have $\mu_1(V) - \mu_2(V) \leq \frac{1}{n-1}$, which is impossible.

Remark 2.5: In nonzero characteristic, the above does not hold as Ein [2] has constructed bundles of rank 2 on $\mathbb{P}^2$ of degree zero whose restrictions to a general $\mathbb{P}^1$ is not semistable. However, since the moduli space of semistable sheaves of rank 2 and degree 0 on $\mathbb{P}^2$ is irreducible [10], one can easily deduce that there exists a nonempty open subset of the moduli space such that the corresponding bundles have semistable reductions.

Section III

Now we sketch a proof of the following, due to Maruyama [9]:

Theorem 3.1: Let V be a semistable vector bundle on a smooth projective variety X. If $\mathrm{rk}E < \dim X$, then for a general $Y \in |\theta_X(1)|$, E/Y is also semistable.

First we need a construction: Let $X \subset \mathbb{P}^n$ and let G_1 and G_2 be the Grassmannians of codimension 1 and codimension 2 linear spaces in $\mathbb{P}^n$, respectively.

Define $\tilde{X}$ by:

$$\tilde{X} = \{(x,V,W) \in X \times G_1 \times G_2 \mid x \in V \supset W\}$$

Define $\tilde{Y}$ by: $\tilde{Y} = \{(x,V,W) \in \tilde{X} / x \in W\}$.
Define a correspondence variety T between G_1 and G_2 by
$T = \{(V,W) \in G_1 \times G_2 / V \supset W\}$.

Then we have:

$$\begin{array}{ccccc} \tilde{Y} & \subset & \tilde{X} & \xrightarrow{q} & T \\ & & \Big\downarrow{\scriptstyle p} & & \\ & & X & & \end{array}$$

Suppose E is a vector bundle on X and suppose $E|X_K$ and $E|Y_K$ are both unsemistable, where K is the quotient field of T.

Let $0 = E_o \subset E_1 \subset \ldots \subset E_m = P^*(V)/\tilde{X}$ and

$0 = F_o \subset F_1 \subset \ldots \subset F_n = P^*(V)/\tilde{Y}$ be the Harder-Narasimhan filtrations of $P(V)/\tilde{X}$ and $P(V)/\tilde{Y}$, respectively. Under the above conditions, we have

Lemma 3.2: If for some i, $1 \le i \le m-1$, $E_i/Y \simeq F_j$, for some j, $1 \le j \le n-1$, then there exists a subsheaf M of E such that $M/\tilde{X} \simeq E_i$. In particular, E is not semistable.

Proof: Let $x \in G_2$ and consider $p_2^{-1}(x)$, where p_1 and p_2 are the two projections: $T \to G_1$ and $T \to G_2$. Then $p_2^{-1}(x)$ is a $\mathbb{P}^1$, imbedded via p_1 in G_1. Put $Z = q^{-1}(\mathbb{P}^1)$. Then Z is the blow-up of X along Y, where Y is the Scheme-theoretic intersection of X with the Codim 2 linear space in $\mathbb{P}^n$ corresponding to x. We have the diagram:

$$\begin{array}{ccccccc} p^{-1}(Y) & = & Y \times \mathbb{P}^1 & \subset & Z & \xrightarrow{q} & \mathbb{P}^1 \\ & & \Big\downarrow & & \Big\downarrow{\scriptstyle p} & & \\ & & Y & \subset & X & & \end{array}$$

Consider E_i/Z, $E_i/Y \times \mathbb{P}^1$ and $F_j/Y \times \mathbb{P}^1$. By assumption, the last two bundles on $Y \times (t)$ are isomorphic for every $t \in \mathbb{P}^1$. If rank $E_i = s$, then E_i/Z gives a map $\sigma : Z \to G_s(E)$, the Grassmannian of rank s subbundles of E. By restriction

σ to $Y \times \mathbb{P}^1$, one sees that is independent of the $\mathbb{P}^1$ factor, hence σ descends to a map $\tilde{\sigma} : X \to G_s(E)$, which defines M. Hence E is not semistable.

Now the proof of Theorem 3.1 follows by induction: Let Y be a hyperplane section of X. If E/Y is not semistable and if $0 = E_o \subset \ldots \subset E_t = E/Y$ is the Harder-Narasimhan filtration then $rkE_i < rkE$, $1 \le i \le t-1$. So far a hyperplane section Z of Y, E_i/Z is semistable $\forall$ i. In particular $0 = E_o/Z \subset \ldots \subset E/Z$ is the Harder-Narasimhan filtration of E/Z. Then by Lemma 3.2 there is a subsheaf $\bar{E}$ of E with $\bar{E}/Y = E_i$. Clearly $\bar{E}$ contradicts the semistability of E.

In [2] Ein proves the following:

Let E be rank 2 stable on $\mathbb{P}^2$ in char $p > 0$. Let $\deg E = -1$ and assume that for a general line $L \subset \mathbb{P}^2$, $E/L = \theta_L(a) + \theta_L(-a-1)$, with $a \ge 0$. Then $a \le \frac{1}{2}[(\frac{4c_2(E)-1}{3})^{1/2} - 1]$. And if $\deg E = 0$ then $a \le (c_2(E)/3)^{1/2}$. He proves this by factoring the map from the correspondence variety to $\mathbb{P}(E)$ by its Frobenius transform. See also [10, Lemma 7.3] and [14, Thm. 8.2] where similar results are proved using the Riemann-Roch Theorem.

REFERENCES

[1] Barth, W.: "Some properties of stable rank-2 vector bundles on $\mathbb{P}^n$", Math. Ann. 226, 125-150 (1977).

[2] Ein, L.: "Stable Vector Bundles on projective spaces in Char. $p > 0$", Math. Ann. 254, 53-72 (1980).

[3] Ein, L., Hartshorne, R., Vogelaar, H.: "Restriction Theorems for Stable Rank 3 Vector Bundles on $\mathbb{P}^n$", Math. Ann. 259, 541-569 (1982).

[4] Elencwajg, Hirschowitz, Schneider: "Les Fibrés Uniforms de Rang au Plus n sur $\mathbb{P}^n(\mathbb{C})$ Sont Ceux Qu'on Croit", (Preprint).

[5] Forster, Hirschowitz, Schneider: "Type De Scindage Généralisé Pour Les Fibrés Stables", (Preprint).

[6] Lang, H.: "On Stable and Uniform Rank - 2 Vector Bundles on $\mathbb{P}^2$ in characteristic P", Manuscripta Math. 29, 11-28 (1979).

[7] Langton, S.: "Valuative criteria for families of vector bundles on algebraic varieties", Ann. Math. 101, 88-110 (1975).

[8] Maruyama, M.: "The Theorem of Grauert-Mullich-Spindler", Math. Ann. 255, 317-333 (1981).

[9] Maruyama, M.: "Boundedness of Semistable Sheaves of Small Ranks", Nagoya Math. J. 78, 65-94 (1980).

[10] Maruyama, M.: "Moduli of Stable Sheaves II", J. Math. Kyoto Univ., 18 (1978).

[11] Mehta, V.B., Ramanathan, A.: "Semistable Sheaves on Projective Varieties and their Restriction to Curves", Math. Ann. 258, 213-224 (1982).

[12] Ramanan, S., Ramanathan, A.: "Some remarks on the Unstability flag", (Preprint).

[13] Weil, A.: "Sur les Critère d'équivalence en géometrie algébrique", Math. Ann. 128, 95-127.

[14] Hartshorne, R.: "Stable Vector Bundles of rank 2 on $\mathbb{P}^3$", Math. Ann. 238, 229-280 (1978).

[15] Schneider, M.: Chernklassen semi-stabilen Vektorraum bündel Vom Rang 3 auf Hyperebenen des Projectiven Raumes. Grelle J. 323, 177-192 (1981).

Acknowledgements: During the preparation of this paper the author was a Visiting Professor at the University of Naples, supported by the Consiglio Nazionale delle Ricerche, Italy. He is grateful to the University of Naples and the C.N.R. for their hospitality.

FONDAZIONE C.I.M.E.
CENTRO INTERNAZIONALE MATEMATICO ESTIVO
INTERNATIONAL MATHEMATICAL SUMMER CENTER

"Complete Intersections"

in the subject of the First 1983 C.I.M.E. Session.

The Session, sponsored by the Consiglio Nazionale delle Ricerche and the Ministero della Pubblica Istruzione, will take place under the scientific direction of Prof. SILVIO GRECO (Politecnico di Torino, Italy) at the Azienda Regionale delle Terme, Acireale (Catania), Italy, ***from June 13 to June 21, 1983.***

Courses

a) ***Complete intersections in affine-algebraic spaces and Stein spaces.*** (8 lectures in English).
Prof. Otto FORSTER (Ludwig-Maximilians-Universität, München, BRD).

1. Estimate of the number of generators of ideals in non-local rings. Proof of the Forster-Eisenbud-Evans conjecture.
2. Estimate of the number of equations necessary to describe algebraic (analytic) sets. Proof of the theorem of Storch-Eisenbud-Evans.
3. The role of the normal bundle.
4. Topological conditions for ideal-theoretical complete intersections in Stein spaces.
5. The Ferrand construction. Set theoretical complete intersections.

References

1. BANICA-FORSTER, Complete intersection in Stein manifolds. Manuscr. Math. 37 (1982), 343-356.
2. EISENBUD-EVANS, Every algebraic set in n-space is the intersection of n hypersurfaces. Inv. Math. 19 (1973), 107-112.
3. FERRAND, Courbes gauches et fibrés de rang 2. CR Acad. Sci. Paris 281 (1975), 345-347.
4. FORSTER, Uber die Anzahl der Erzeugenden eines Ideals in einem Noetherschen Ring. Math. Z. 84 (1964), 80-87.
5. FORSTER-RAMSPOTT, Analytische Modulgarben und Endromisbündel. Inv. Math. 2 (1966), 145-170.
6. KUNZ, Einführung in die kommutative Algebra und algebraische Geometrie, Kap. V., Vieweg 1980.
7. MOHAN KUMAR, On two conjectures about polynomial rings. Inv. Math. 46 (1978), 225-236.
8. SATHAYE, On the Forster-Eisenbud-Evans conjecture. Inv. Math. 46 (1978), 211-224.
9. SCHNEIDER, Vollständige, fast-vollständige und mengentheoretischvollständige Durchschnitte in Steinschen Mannigfaltigkeiten. Math. Ann. 260 (1982), 151-174.
10. STORCH, Bemerkung zu einem Satz von M. Kneser. Arch. Math. 23 (1972), 403-404.
11. SWAN, The number of generators of a module. Math. Z. 102 (1967). 318-322.
12. SZPIRO, Equations defining space curves. Tata Institute Bombay, Springer 1979.

b) ***Work of Zak and others on the geometry of projective space.*** (8 lectures in English).
Prof. Robert LAZARSFELD (Harvard University, USA).

A conjecture of Hartshorne, to the effect that any smooth subvariety of sufficienty small codimension in projective space must be a complete intersection, has sparked a considerable body of work over the past decade. We will survey some of these results, focusing on Zak's recent solution of a related problem of Hartshorne's on linear normality. Specifically, the course will be organized as follows:

1. Historical Introduction; theorems of Barth, Fulton-Hansen, et. al.
2. Work of Zak.
3. Further results; open problems.

References:

1. R. HARTSHORNE, Varieties of small codimension in projective space, Bull. A.M.S. 80 (1974), 1017-1032.
2. W. FULTON and R. LAZARSFELD, Connectivity and its applications in algebraic geometry, in Libgober and Wagreich (eds), Algebraic geometry, Proceedings, Chicago Circle (1980), Lecture notes in math. no. 862, Springer Verlag.

c) ***Complete intersections in weighted projective spaces.*** (4 lectures in English).
Prof. Lorenzo ROBBIANO (Università di Genova, Italy).

The purpose of this course is to give a brief account of some results relating classical theorems on complete intersections in projective spaces to new results in weighted projective spaces.

The first part will treat some basic facts on weighted projective spaces, while the second one will be concerned with more specialized facts, such as Lefschetz-type theorems. In particular, problems of factoriality and semifactoriality will be studied.

Basic references

1. C. DELORME, Espaces projectifs anisotropes, Bull. Soc. Math. France 103 (1975).
2. M. DEMAZURE, Anneaux gradués normaux, in Séminaire Demazure-Giraud-Teissier, Singularités des surfaces, Ecole Polytechnique 1979.
3. I. DOLGACHEV, Weighted projective varieties. Mimeographed notes. Moscow State University 1975/76.
4. R.M. FOSSUM, The divisor class group of a Krull domain, Ergeb. Math. Grenz. Bd. 74, Springer Berlin 1973.
5. S. MORI, On a generalization of complete intersections, J. Math. Kyoto Univ. 15 (1975).

d) ***On set-theoretic complete intersections.*** (4 lectures in English).
Prof. Giuseppe VALLA (Università di Genova, Italy).

The aim of this course is to give a comprehensive approach to some of the research frontiers in the topic of algebraic varieties which are set-theoretic complete intersections.

Focusing on the special case of affine or projective algebraic curves over a field of characteristic zero, the course will develop to include the most important and recent results on this subject, such as the theorems, given by D. Ferrand and M. Kumar, on affine curves which are locally complete intersections.

The final part of the course will be devoted to make some hints at the case of projective space curves.

Reference

1. J.P. SERRE, Sur le modules projectifs, Sem. Dubreil-Pisot 14 (1960/61).
2. P. MURTHY, Complete intersections, Conference on Commutative Algebra 1975, Queen's Unviersity, 196-211.
3. M. KUMAR, On two conjectures about polynomial rings, Inv. Math. 46 (1978), 225-236.

Seminars

A number of seminars and special lectures will be offered during the Session.

FONDAZIONE C.I.M.E.
CENTRO INTERNAZIONALE MATEMATICO ESTIVO
INTERNATIONAL MATHEMATICAL SUMMER CENTER

"Bifurcation Theory and Applications"

in the subject of the Second 1983 C.I.M.E. Session.

The Session, sponsored by the Consiglio Nazionale delle Ricerche and the Ministero della Pubblica Istruzione, will take place under the scientific direction of Prof. LUIGI SALVADORI (Università di Trento, Italy) at Villa «La Querceta», Montecatini Terme (Pistoia), Italy, ***from June 24 to July 2, 1983.***

Courses

a) ***Bifurcation Phenomena in Biomathematics***. (6 lectures in English).
Prof. Stavros BUSENBERG (Harvey Mudd College, USA).

Lecture 1: Origins of bifurcation problems in biomathematics.
Nonlinear interactions in population dynamics, nerve pulse propagation, cell growth and morphogenesis.

Lecture 2: Bifurcation and stability in models with monotone properties.
Global bifurcation and stability of constant, periodic and almost periodic solutions. Applications to epidemic and other population models.

Lecture 3: Hopf type bifurcation.
Models in population dynamics and metabolic control with Hopf bifurcations. Periodic, quasiperiodic and chaotic behavior.

Lecture 4: Linear and nonlinear diffusion.
Spatial diffusion and pattern formation. Chemotaxis, strain guided diffusion and morphogenesis.

Lecture 5: Separable age-dependent processes.
A method for decomposing the equations of age-structured processes. Bifurcation phenomena in age-dependent population and cell growth models.

Lecture 6: Diffusion in age-dependent processes.
Spatial diffusion in age-dependent population and cell growth models. Bifurcation of spatially heterogeneous solutions and the development of spatial structure.

References

General background: Mathematics of Biology, M. Iannelli editor, CIME ciclo 1979, Liguori, Napoli (1981). Hoppensteadt, F., Mathematical Theory of Population, Demographics, Genetics and Epidemics, SIAM, Philadelphia (1975).

Population and epidemic models: Busenberg, S. and Cooke, K., «The effect of integral conditions in certain equations modelling epidemics and population growth», J. Math. Biol. 10 (1980), 13-22. Busenberg, S. and Cooke, K., «Models of vertically trasmitted diseases with sequential continuous dynamics», in Nonlinear Phenomena in Mathematical Science, V. Lakshmikantham, editor, Academic Press, New York (1982). Lajmanovich, A. and Yorke, J., «A deterministic model for gonorrhea in a nonhomogeneous population», Math, Biosc. 28 (1976), 221-236.

Diffusion an age-dependence: Busenberg, S. and Travis, C., «Epidemic models with spatial spread due to population migration», J. Math. Biol. (1983) (in press). Busenberg S. and Iannelli, M., «A class of nonlinear diffusion problems in age-dependent population dynamics», Nonlinear Analysis MTA, (1983) (in press). Okubo, A., Diffusion and Ecological Problems: Mathematical Models, Springer Verlag, New York (1980).

b) ***Bifurcation of periodic solutions near equilibria of Hamiltonian systems***. (6 lectures in English).
Prof. I.J. DUISTERMAAT (State University of Utrecht, NL).

A short outline of the content

The problem of finding periodic solutions is formulated as an equation in the loop space. Using Lyapunov-Schmidt reduction this is equivalent to finding zeros of a vectorfield in a finite dimensional space, with a built-in circle invariance. Variants of this procedure work for discrete dynamical systems and for finding homoclinic orbits. If the original system is Hamiltonian then the reduced problem amounts to finding critical points of a circle-invariant function. Near equilibrium points the problem is solved approximately up to any order using Birkhoff normal forms. If only two degrees of freedom are in resonance then, in the generic case, one can bring the equations into an exact normal form, using Wasserman's group invariant version of Mather's theory. If the system itself has two degrees of freedom then this also gives useful information about the other solutions near the equilibrium points. The course will be concluded with a discussion of the situation when more than two degrees of freedom are in resonance.

References

1. D.S. SCHMIDT, Periodic solutions near a reasonant equilibrium of a Hamiltonian system, Celestial Mechanics 9 (1974), 81-103.
2. J. MOSER, Periodic orbits near an equilibrium and a theorem by Alan Weinstein, Comm. Pure Appl. Math. 29 (1976), 727-747.
3. G. WASSERMAN, Classification of singularities with compact abelian symmetry, Regensburger Math. Schriften 1, 1977.

c) ***Topics in Bifurcation Theory***. (6 lectures in English).
Prof. Jack K. HALE (Brown University, USA).

The topics include:

Bifurcation from an equilibrium point with one zero or two pure imaginary roots, relations between the bifurcation function and the center manifold, and the extent to which the theory is valid in infinite dimensions. Nonlocal results and some codimension two bifurcations in R^2 and the role of symmetry. Generic theory, dynamic behavior and stable equilibria in a parabolic equation. Nonlinear oscillations and chaotic behavior in functional differential equations.

The basic references are:

1. S.N. CHOW and J.K. HALE, Methods of Bifurcation Theory, Grundlehren der Math. Wiss. 251, Springer-Verlag, 1982.
2. J.K. HALE, Topics in Dynamic Bifurcation Theory, NSF-CBMS Lectures 27, Am. Math. Soc., Providence, R.I. 1981.

d) ***Bifurcation and transition to turbulence in hydrodynamics***. (6 lectures in English).
Prof. Gérard IOOSS (Université de Nice, F).

Outline of the contents:

Physical motivation - Experimental results.
Navier-Stokes equations as a dynamical system, regularity properties of the solution. Poincaré map.
Specific examples: Taylor problem, plane Bénard problem.
Bifurcations which break symmetries, rotating waves, quasiperiodic solutions, frequency lockings.
Routes for transition to turbulence. Conjectures, open problems.

Basic literature references:

1. V.I. ARNOLD, Chapitres supplémentaires de la théorie des équations différentielles ordinaires, ed. MIR, Moscou 1980.
2. S.N. CHOW, J.K. HALE, Methods of bifurcation theory, Springer Verlag, 1982.
3. G. IOOSS, Arch. Rat. Mech. Anal., 64, 4 (1977), 339-369.
4. G. IOOSS, Bifurcation of maps and applications, North Holland Math. Stud. 36, 1979.
5. D.D. JOSEPH, Stability of fluid motions, vol. I and II, Springer Tracts in Phil., vol. 27, 28, 1976.
6. T. KATO, Perturbation theory for linear operators, Springer Verlag, 1966.
7. O.A. LADYZENSKAYA, The mathematical theory of viscous incompressible flow, Gordon and Breach, 1969.
8. J. MARSDEN, M. Mc CRACKEN, The Hopf bifurcation and its applications, Math. Applied Sciences, 1, Springer Verlag 1976.

Seminars

A number of seminars and special lectures will be offered during the Session.

FONDAZIONE C.I.M.E.
CENTRO INTERNAZIONALE MATEMATICO ESTIVO
INTERNATIONAL MATHEMATICAL SUMMER CENTER

"Numerical Methods in Fluid Dynamics"

in the subject of the Third 1983 C.I.M.E. Session.

The Session, sponsored by the Consiglio Nazionale delle Ricerche and the Ministero della Pubblica Istruzione, will take place under the scientific direction of Prof. FRANCO BREZZI (Università di Pavia, Italy) at «Villa Olmo», Como, Italy, ***from July 4 to July 12, 1983.***

Courses

a) ***Finite Elements Method for Compressible and Incompressible Fluids.*** (6 lectures in English).
Prof. Roland GLOWINSKI (INRIA, France).

1. Finite Elements Method for the Stokes problem.
2. Nonlinear least-square and applications to fluid flow problems.
3. Alternating directions methods for Navier-Stokes equations.
4. Upwinding methods for transonic flows.

References

1. GIRAULT, RAVIART, Finite Element Approximation of Navier-Stokes equations. Lecture Notes in Math. n. 749 (1979), Springer.
2. GALLAGHER, NORRIE, ODEN, ZIENKIEWICZ (Eds.), Finite Elements in Fluids. Vol. IV. J. Wiley, 1982.
3. GLOWINSKI, Numerical Methods in Nonlinear Variational Problems. Cap. VII, Springer, 1983.

b) ***Spectral methods for partial differentiation equations of fluid dynamics.*** (6 lectures in English).
Prof. David GOTTLIEB (NASA, USA).

Lecture 1 : Presentation of spectral methods - Fourier Chebyshev and others, survey of approximation results.
Lecture 2+3: Stability and convergence of spectral methods for parabolic and hyperbolic P.D.E.'s.
Lecture 4 : Time marching and iterative techniques.
Lecture 5 : Application - incompressible flows.
Lecture 6 : Application - compressible flows.

Literature

1. D. GOTTLIEB & S.A. ORSZAG, Numerical Analysis of Spectral Methods, Theory and Applications C.B.M.S.-S.I.A.M. No. 26, 1977.
2. B. MERCIER, Analyse numérique des Méthodes Spectrales. Note CEA-N-2278 Commissariat à l'Energie Atomique, Centre d'études de Limiel.

c) ***Transonic flow calculations for aircrafts.*** (6 lectures in English).
Prof. Antony JAMESON (Princeton University, USA)

— Review of mathematical models
— Potential flow methods
— Multigrid acceleration
— Solution of the Euler equations in 2 and 3 dimensions.

References

1. K.W. MORTON, R.P. RICHTMYER, Difference methods for initial value problems, New York, 1967.
2. A. BRANDT, Math. Comp. 31 (1977).

3. A. JAMESON, Comm. Pure & Appl. Math., 27 (1974).
4. A. JAMESON, «Steady state solution of Euler equation for Transonic Flow» in «Transonic, shock, and multidimensional flows: Advances in scientific computing», R.E. Meyer ed., Academic Press 1982.

d) ***An Analysis of Particle Methods***. (6 lectures in English).
Prof. P.A. RAVIART (Université P. et M. Curie, Paris).

By particle methods, one usually means numerical methods where some dependent variables of the problem are approximated by a sum of delta functions. This course intends to provide the mathematical basis of these methods which play an increasing role in Fluid Mechanics and in Physics. The following topics will be discussed:

1. Particle approximation of linear hyperbolic equations
2. Numerical approximation of Euler equations in two and three dimensions by vortex and vortex in cell methods.
3. Particle approximation of Vlasov-Poisson equations in plasma physics.

References

1. J.T. BEALE & A. MAJDA, Vortex methods I: Convergence in three dimensions, Math. Comp. 32 (1982), 1-27.
2. J.T. BEALE & A. MAJDA, Vortex methods II: Higher order accuracy in two and three dimensions, Math. Comp. 32 (1982), 29-52.
3. G.H. COTTET, Méthodes particulaires pour l'équation d'Euler dans le plan, Thèse de 3ème cycle, Université Pierre & Marie Curie, Paris 1982.
4. G.H. COTTET & P.A. RAVIART, Particle methods for the one-dimensional Vlasov-Poisson equations, rapport interne 82027, Laboratoire d'Analyse Numérique, Université Pierre & Marie Curie, Paris (to appear in SIAM J. Num. Anal.).
5. R.W. HOCKNEY & J.W. EASTWOOD, Computer simulation using particles, McGraw-Hill, New York, 1981.
6. A. LEONARD, Vortex methods for flow simulation, J. Comp. Physics, 37 (1980), 289-335.

Seminars

A number of seminars and special lectures will be offered during the Session.